HEADLESS MALES MAKE GREAT LOVERS

HEADLESS MALES MAKE

MARTY CRUMP

With illustrations by Alan Crump

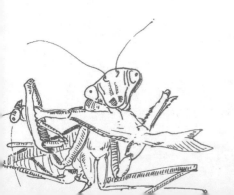

GREAT L♥VERS

& OTHER UNUSUAL NATURAL HISTORIES

The University of Chicago Press CHICAGO & LONDON

MARTY CRUMP is an adjunct professor in the Department
of Biological Sciences at Northern Arizona University. She is the
author of *In Search of the Golden Frog* and *Amphibians, Reptiles, and
Their Conservation* and a coauthor of *Herpetology*.

The University of Chicago Press, Chicago 60637
The University of Chicago Press, Ltd., London
Text © 2005 by The University of Chicago
Illustrations © 2005 by Alan Crump

14 13 12 11 10 09 08 07 06 05   1 2 3 4 5

ISBN: 0-226-12199-2

Library of Congress Cataloging-in-Publication Data
Crump, Martha L.
    Headless males make great lovers: and other unusual natural
histories / Marty Crump; with illustrations by Alan Crump.
        p.   cm.
    Includes bibliographical references.
    ISBN 0-226-12199-2 (alk. paper)
    1. Animal behavior.   I. Title.
    QL751.C8825 2005
    590—dc22

                                            2005007592

To the memory of Robert M. Crump and to Ruth M. Crump,

who shared their love of natural history with us

Contents

Preface

ZOOLOGISTS ARE a curious lot. They poke and probe animals to understand how they work. They enclose everything from ants to anteaters in chambers and measure water loss or respiratory rate. They count scales and teeth on preserved specimens in hopes of discovering new species. Some zoologists watch animals to describe their natural history: how they choose their mates, whether or not and what type of care they provide for their young, what food they eat and how they get it, how they defend themselves, and how they communicate. I'm one of the latter group.

This book celebrates animal diversity. I've purposely focused on bizarre "gee whiz" stories. Some are based on my own research or that of friends and colleagues; others come from scientists whom I don't know personally. The animals run the gamut from sponges to mammals, and they include examples from all over the world. The animal natural histories I've selected are an eclectic assortment, grouped into five fundamental themes: mating games, parental care, food and feeding, defense, and communication. They are a tiny sampling of the fascinating animal behaviors that naturalists have described.

My goal in writing this book was to convert scientific observations, some of which lie buried in the technical literature, into a readable form for a general audience. If I can whet the appetite of a few budding naturalists and increase curious laymen's appreciation for natural history, then I will have been successful.

Scientists writing for a general audience sometimes straddle a fine line. We're either too technical, stilted, and dry, or in our enthusiasm to excite the reader, we come across as teleological (assuming animals behave with a conscious design or purpose in mind) or anthropomorphic (attributing human characteristics to animals). If in my effort to engage the reader I sometimes stray from the center line, please know that I do not suggest that animals behave with a conscious goal in mind or that they react to situations as a human would. I'm simply tripping over my own enthusiasm.

Although I've avoided scientific jargon, I have defined some technical words in the text. Occasionally I use order, family, or other levels of scientific classification. For readers unfamiliar with these terms, please see the appendix.

The discussions of various animal behaviors are *not* meant to be rigorous scientific reviews, nor are they meant to be exhaustive. I have opted for breadth rather than depth. I encourage readers seeking more information to consult the references, which range from general articles written for the nonbiologist to textbooks and technical scientific papers.

Join me in celebrating the diversity of nature with some amazing animal behaviors, from various insects and birds that offer food as bribes for sex to sac-winged bats that blend their own aphrodisiacs from urine and smelly glandular secretions.

Acknowledgments

WE THANK OUR editor, Christie Henry, for her insight and enthusiasm throughout this project. In addition, we thank Erin DeWitt of the University of Chicago Press for expert copyediting.

Members of Marty's weekly writing group—Martha Blue, Carol Eastman, Kay Jordan, Carol Maxwell, Mona Mort, and Marilyn Taylor—not only tolerated a scientist in their midst; they also offered support and guidance from their nonbiologist perspectives. Karen Feinsinger and Judy Hendrickson provided additional helpful comments and suggestions on the text. Peter Feinsinger read and commented on the entire manuscript not once but twice; Marty thanks him for his suggestions, support, and encouragement.

Alan thanks Irma Crump for her thoughtful and creative suggestions, support, and patience when he was lost in the inkwell.

We applaud and are grateful to the many naturalists and scientists who recorded their fascinating observations of unusual animal behavior.

1 Ain't Love Grand!

PEOPLE OFTEN ASSUME that the evolutionary goal of copulating cats or mating monkeys is "to perpetuate the species." If this were true, males and females should approach mating cooperatively. Why then do we see so much hostility between the sexes? The goal is not to perpetuate the species but rather to perpetuate one's own genes. From the individual's standpoint, the goal is to produce as many offspring (and grandchildren, great-grandchildren, etc.) as possible. Individuals that leave behind the greatest number of descendants are the most "fit," and males and females differ in how they best increase their fitness.

The typical male strategy is to mate with as many females as possible, which often leads to fierce competition among males and requires elaborate displays to woo females. On the other hand, a female increases her fitness by mating with the "best" male available; she's choosy and rejects most suitors. Most females in a population mate, but often only a small fraction of the males mate.

The contrast in the ways males and females increase their fitness results from differences between their gametes (sex cells). A female produces relatively few large eggs that contain the resources needed for early development. Before a female's eggs are even fertilized, she has already made a substantial investment in each sex cell. In contrast, a male produces vast amounts of tiny mobile sperm. Consider humans. An average woman produces about 350 to 400 mature eggs during her lifetime. Each ejaculate from a healthy man consists of from 2 to 6 milliliters of sperm, each milliliter containing about 100 million sperm. For animals in general, a female's reproductive potential is much lower than a male's because she produces many fewer gametes. A female can hardly afford to waste her sex cells; a male can sow widely and often.

Throughout this chapter, we'll look at various ways animals attempt to increase their fitness: (1) competition among males, from bull elephant seals defending harems to male tiger salamanders tricking other males into depositing (and wasting) their packets of sperm; (2) elaborate male courtship to woo females, from nuzzling and dancing spiders to bugling elk; and (3) female choice of males, sometimes based on unusual criteria such as a male's fishing ability.

Reproduction can be complicated. Instead of attracting females on their own, some males sneak into a harem and seize a sexual moment or they mimic females to get near a couple and then attempt to mate. The mating game can be dangerous. Females of some praying mantids and spiders eat their suitors before, during, or after copulation. Males of many species display unique ways to avoid being cannibalized, including kidnapping and trickery. Sometimes the female must be bribed into mating; many male insects and birds offer females nuptial gifts of food in return for sex.

Not all animals are either females that produce eggs or males that produce sperm. Some are hermaphrodites. Simultaneous hermaphrodites produce eggs and sperm at the same time. Sequential hermaphrodites begin as one sex and produce one type of gamete; at a later stage, they switch over and produce the other type. Some of these animals get fairly kinky in their sexual behavior, but they have the same goal as nonhermaphrodites: to leave behind as many descendants as possible.

RAMPANT MACHISMO

When the beachmaster [male elephant seal] is angered by a serious challenge, he thunders across the sand, humping and heaving his huge body with surprising speed and taking no notice whatever of what lies in his way.

DAVID ATTENBOROUGH, *The Trials of Life*

Loud snorts erupted as I ambled toward the beach on Península Valdés, Argentina. Around the bend lay hundreds of giant brown or gray four-ton sausages parked from one end of the beach to the other. Binoculars revealed the sausages to be southern elephant seals.

Some slept, using one another for pillows. Others lazily rolled over from time to time. One balanced on his front flippers, humped his body, and propelled himself forward, like a giant inchworm crossing the sand. A smaller individual lumbered after another, in what appeared to be a sausage race. As I focused on one particularly large, ponderous pile of blubber, he threw back his head, proboscis dangling down in his open mouth, and roared.

Related to walruses and sea lions, elephant seals (sometimes called sea elephants) can reach twenty feet and weigh four tons. Both sexes spend much of the year in the ocean but come ashore to islands or remote continental shores to breed. Their unique body shape is well designed to reduce drag while swimming. Their tough skin, large body size, and the trunk-shaped drooping nose of males (bulls) are reminiscent of elephants.

Elephant seals display a reproductive system described as "rampant machismo." Males increase their reproductive success by preventing other males from mating. Bulls arrive at the rookeries (breeding grounds) before females and struggle to establish dominance. Only the highest-ranking, most dominant males will gain access to females. Bulls roar and rush toward one another. They butt chests. They rip into each other's tough skin with their teeth, cutting noses and extracting chunks of flesh and blubber. One champion male emerges victorious. Eventually the females (cows) arrive, pregnant from last season's mating. They seek each other out for safety from the aggressive males, forming a harem, which the head male attempts to monopolize. The second-ranked male tries to keep all other males away from the cows but defers to the harem master. And so it goes for each male, according to his rank. A bull's social rank is constantly threatened, so he must be vigilant.

Harem size varies. In groups containing fewer than 50 females, usually one alpha bull does all the mating. With a larger harem—of, say, 250 to 300 females—as many as 10 to 15 males can mate simply because one harem master can't be everywhere at once guarding against intruders. In a large harem, the most dominant bull patrols and rules one area. The second-most dominant male occupies and defends a smaller area, and the third-ranked male an even smaller area. Low-ranking males operate at the harem edges, but even they must defend their few females from intruders.

Only the strongest bulls become harem masters, and only about 10 percent of the males fertilize virtually all the females. The higher a male's rank, the more likely he can mate without other bulls interfering. Life isn't easy for harem masters, however. They can't eat for over a hundred days. If they venture out to sea to eat, other males will certainly take over. Stress is so great that these dominant males frequently die the year following their reign, and they rarely maintain their master status for more than three consecutive years.

Low-ranking males lead a different life. They attempt to mate with cows but are usually attacked by higher-ranking males either before or during copulation. Many try to sneak copulations when the dominant males are fighting, sleeping, or mating. Despite valiant efforts, many males die without ever having mated.

It would seem that cows can't choose their mates, but they actually have some say—by influencing the outcome of competition between males. A female will sometimes bellow loudly when a low-status bull attempts to copulate with her. Hearing the protests, a dominant bull will chase away the low-ranking male and likely mate with her. By inciting males to compete for her and mating only with the most dominant bulls, a female is more likely to have sons that will inherit macho genes associated with male strength and dominance.

About six days after the females arrive on the rookery, each gives birth to a single pup from last year's mating. Going without food, the mothers nurse their pups for about twenty-eight days and then return to sea. At this point, their young are weaned.

Although females are receptive to males only during the last few days of nursing their pups, bulls attempt to mate with any and all cows regardless of a female's reproductive condition—pregnant, giving birth, or soon afterward. With their huge hulking bodies, at more than twice and sometimes ten times the size of females, males overpower their mates. Bulls restrain the females by clasping them with their front flippers and by biting them in the neck. Pups naturally want to stay close to their mothers. Those that get in the way during a male's determined but clumsy attempt to mate may be trampled to death.

Cows respond to males in one of two ways. They may actively protest by vocalizing loudly, whipping their bodies from side to side, struggling to get away, whacking the males with their flippers, dashing sand in their faces, and nipping their aggressive suitors' necks. But if a female is receptive to a particular male's advances, and if the time is right in her cycle, they mate.

THE ELEPHANT SEAL mating system may seem peculiar, but quite a few mammals (and some birds, fishes, and other animals) form and defend harems. We'll look at a sampling of mammal harems here. A few other animals that form harems are mentioned elsewhere in this chapter.

In most animals, males compete for access to females, and the most dominant males are the most successful in leaving behind offspring. If females group together for whatever reason, a male can exert dominance and defend his access to those females. As we saw, female elephant seals group together for protection from the aggressive bulls. In some other animals, females gather where food is distributed in patches; in those cases, an individual male may be able to monopolize the group of females.

Female impalas (African antelope) aggregate while browsing in wooded grasslands, presumably as a defense against lions and cheetahs. There is power in numbers: the more pairs of eyes and ears on alert, the more likely each female can escape. Male impalas defend territories. A male with a territory that contains good grazing and cover conditions is more likely to have a group of females wander into and stay in his territory than a male who presides over barren land. Although a territory owner has "rights" to the females within his domain, he must fight against intruder males, who don't have their own harems and would like to sneak in and steal sex.

Bull North American elk likewise guard harems, and they are captivating to watch. At dusk in late September, my brother, Alan, saw about forty elk stream onto the Estes Park, Colorado, golf course. He parked next to a sign: "Unauthorized Entry to the Golf Course Prohibited" (obviously the elk were not intimidated) and watched for the next forty-five minutes. One magnificent bull, "the Big Guy," was the obvious harem master. He bellowed loudly and galloped from one end of the golf course to the other frantically defending his harem from five other males. He locked horns with one bull, but three intruders mated with the master's cows simply because the Big Guy couldn't be everywhere at once. Alan recalled, "The Big Guy would be off at one end of the golf course, at about the third hole, chasing a rival, and another bull would seize the moment at the fourteenth hole. The Big Guy was in constant motion, as were all the elk, so it was like watching a hypnotic dance—fluid, dramatic, and primeval."

Female elk choose the "best" males available to father their offspring— the best being large dominant bulls. To a female, a male with large antlers sprouting many points oozes competence and good health. He has demonstrated superior ability to eat well and convert the excess nutrients to antler tissue. A male that eats only enough to maintain his body cannot grow large showy antlers. Large males have also successfully escaped predators. Presumably the traits of competence and survival ability are passed on to offspring. Since cows choose their mates, males compete among themselves. During early autumn, males advertise their dominance and attempt to attract a harem of twelve to sixty females.

Bulls use loud, deep calls to attract females. Toward the end of a bugle, a male sprays urine onto his belly, chest, and neck. The more he bugles, the more urine he sprays onto himself, and the stronger the smell—his own blend of cologne. Sometimes a bull uses his front legs and antlers to dig a wallow in the ground, into which he lowers himself. There he bugles and sprays urine and rolls around in the soil, caking his body with smelly mud. Bulls engage in another endearing attention-grabbing behavior: they "horn" vegetation as a dominance display. A bull scrapes the trunk of a tree with his antlers or teeth, thrashing about as he strips the tree's bark.

Once a male has formed a harem, he must block hesitant females from escaping. He herds a wandering cow by circling her and yelping. A cow generally returns to the group in a submissive posture, with her head and neck lowered. If she doesn't cooperate, the bull may bob his antlers at her as a threat, or even gore her.

In addition to controlling the cows, a bull must defend his harem from other bulls. He does this by asserting his dominance in the same way he attracted females: bugling, urine spraying, wallowing, and horning. Fighting, a last resort because of the danger of injury, includes head-on clashes with antlers lowered, locking of antlers, and wrestling. If one bull stumbles, the other may gore him.

The list of mammals that form harems is impressive—northern fur seals, some baboons, several species of bats, Grant's gazelles, blue wildebeests, pronghorn, Eurasian red deer, camels, vicuñas, some zebras, and musk oxen to name but a few. Even humans. Consider the harems of the Muslim world.

DURING THE 1500S and 1600s, the Ottoman Empire controlled what is now Turkey, parts of northern Africa, southwestern Asia, and southeastern Europe. Ottomans were Muslims, and the Qur'an, the holy book of Islam, allows a man four wives and an unspecified number of concubines. Muslim women during the Ottoman Empire were required to wear veils over their faces in public to protect men from temptation, and they usually lived in a separate section of the home called a harem (from the Arabic word *harīm*, meaning "forbidden"). The typical harem of a well-to-do man included the four permissible wives, one or two concubines, and a couple of servants.

Turkish rulers, or sultans, had large harems. Highest in rank was the sultan's mother. She governed the harem. Next were the women who had borne children to the sultan, along with his personal favorites, even if they had not yet borne his heirs. Lower down were servants who carried out important duties, and lowest in rank were the girls bought in slave markets or kidnapped by pirates. The system was dynamic, with the lowest-ranking girls able to work their way up through the hierarchy. Eunuchs guarded these large harems to prevent any carrying on behind the sultan's back.

In contrast to the way bull elephant seals interact with cows, the Qur'an requires a man to treat his harem properly. Although a man commanded respect and obedience as head of the household, to enter the harem he had to behave with moderation and appropriate etiquette. While a harem may have been the site of erotic pleasures, it was also a place where women could think, educate their children, and socialize with each other.

The Ottomans were defeated during World War I, and their empire dissolved four years after the war ended. Sweeping social reforms included outlawing polygamy. In some places women no longer had to cover their faces with veils, they received the right to vote, and men and women were finally declared equal.

In most parts of the world today, national law prohibits men from having more than one wife at a time. There are exceptions, however, such as Saudi Arabia and Afghanistan, where wealthy men still have four wives and are masters of harems. Only wealthy men can afford harems, because Islamic law requires a man to treat all his wives equally, not only in sexual relationships, but also in financial support.

THERE ARE TWO obvious differences between human harems and those of other animals. First, groups of women don't form independently, then arrive as a group on land controlled by a dominant man. Grouping is usually imposed on the women, in contrast to harems of other animals. Second, non-harem-holding men don't routinely attempt surreptitious matings with women in harems.

Meanwhile, back at Península Valdés, bulls continue to roar, threaten, and fight, competing for the right to dominate harems. And low-ranking males sneak matings whenever the harem master isn't looking. If the master catches on, he'll hump and heave his ponderous pile of blubber across the beach, and the low-ranking male had best bow out.

COME UP AND SEE MY ETCHINGS

The cold-eyed scrutiny of generations of females [bowerbirds] has left males chained to their bowers, like housewives to their ironing boards. There they must work to improve and guard their handiwork from ruthless competitors, and wait eagerly to produce on demand an elaborate courtship performance, only to watch female after female slide away to visit rival males. A difficult life, and perhaps one with a message for those persons afflicted with an excess of machismo.

JOHN ALCOCK, *The Kookaburras' Song*

Says the budding artist to the young woman he's met at a party, "Come up and see my etchings." "C'mon over and listen to my CDs," says the punk rocker to a teenaged girl. And so it is with male bowerbirds from Australia and New Guinea, which lure prospective mates into their boudoirs with trinkets.

Humans and bowerbirds are alike in that males try to impress females with their accumulated wealth and artistry. Males of most other animals rely on their strength, physical skill, size, or some aspect of attractiveness. Whereas many male birds depend on their good looks to attract females, bowerbirds march to a different drummer.

Bowerbirds range in size from about seven to fifteen inches. Of the eighteen bowerbird species, males of fifteen clear away areas of ground and construct impressive structures of grass, twigs, and moss called bowers. The birds use their bowers as sites for courtship displays. Depending on the species, a male builds one of three main types of bowers. A "platform bower" is simply a mat of trampled ferns, grasses, lichens, and leaves. A "maypole bower" resembles a tepee made of twigs, interlocking around a sapling, with a front opening. An "avenue bower" consists of two open-ended parallel walls of twigs, arched toward the center, with a long avenue running the length. Males spend weeks or months constructing their bowers, which they'll use during a breeding season of several months.

To further entice prospective mates, male bowerbirds decorate their bowers. Some species are conservative in their decor. Others lavishly adorn the bower entrance with flowers, shells, feathers, berries, bits of colored cloth, and just about any unusual object available: iridescent beetle wings, bottle caps, bits of tin foil, ribbons, snake skins, spiderwebs, and the skulls of small birds and mammals. Even within a species, no two bowers are alike. Different individuals have different ideas on what will impress a female.

The bower entrance serves as a dance pavilion, where the male courts and attempts to lure prospective lovers inside his bachelor's quarters. Mating occurs inside the bower, and the female constructs her simple nest elsewhere. Unlike many birds, in most species of bowerbirds only the females incubate and feed the chicks. Work with satin bowerbirds suggests that nest predation is so high that only about 25 percent of females can rear their chicks. Perhaps since nest predators are so common, two adults caring for the young would create more commotion and would not improve survival rate of the chicks.

Females usually mate only once during a breeding season, so they're exceedingly choosy and spend hours shopping around for the "best" bower. Males mate with as many females as they can attract, and those with the most elaborate bowers and the most spectacular courtship behavior are the most successful. Some males never mate; others mate twenty or thirty times in a season. The bachelor pad really makes a difference.

Why is bowerbird courtship behavior so different from that of other birds? Most bowerbirds are fairly drab. They can't rely on extravagant plumes and flamboyant patches of color to impress females. They build and decorate bowers instead, and it's the bower that impresses the female. In support of this idea, males of the dullest-colored species build and decorate the most elaborate bowers, whereas males of the more colorful species build and decorate the simplest bowers. The latter can attract females by their looks and rely less on bower decoration.

Each bowerbird species has its own story to tell.

ARCHBOLD'S BOWERBIRDS FROM New Guinea are dark gray or black; males in some populations have golden crests and golden feathers on their foreheads. The male builds a simple platform bower three to eight feet in diameter and decorates it with piles of snail shells, beetle wings, and honey-colored resin chips. If he can find blue berries or dead centipedes, he'll use those also. Or better yet, the elegant blue plume of the King of Saxon bird of paradise. When a female arrives at the bower, the male flattens himself and crawls toward her, flapping his wings like a nestling begging to be fed, uttering churring and rasping sounds while holding a small stick crosswise in his bill. If the female is won over by his groveling (and dexterity with the stick?), the pair mates. Often, though, the female rejects the male for the time being and continues checking out his neighbors.

Male brown gardener bowerbirds, olive-brown birds from New Guinea, construct maypole bowers resembling conical huts with gardens. The male piles bits of moss around the base of a small sapling. Using the trunk as the central pillar, he then places twigs in tepee fashion, leaving an opening for the entrance. The end result can be seven feet wide and four feet high. Next, he lays a carpet of green moss at the hut's entrance. Finally, he brings flowers and fruits of various colors and arranges those on the moss. He may add bits of mushrooms and beetle wings. He carefully tends his garden, sometimes arranging the objects by color: yellow fruits in one area, blue in another, red in another. As fruits and flowers fade, the male replaces them. If dead leaves or other debris land on the garden, he fastidiously removes them. When a female approaches, the male thrashes his wings and produces whistles, grunts, catlike meows, and rapping and ticking sounds. If the female decides this is her Prince Charming, the pair disappears into the hut. More likely than not, she bypasses the male and his bower and moves on to one of his competitors.

Spotted bowerbirds from Australia are brown with rufous or buff spots. Males have a vivid patch of lilac pink feathers on the neck; females lack this colorful patch or have only a small one. The male constructs an avenue-type bower of two parallel walls of sticks and grasses, built on a foundation of sticks, with a dance platform at each end. A bower, including the dance platforms, can reach seven feet in length.

A male spotted bowerbird individualizes his bower with sheep and rabbit bones, pebbles, eggshells, snail shells, seed pods, pinecones, green berries, and bits of green, amber, or white glass. These birds seem to favor white, gray, green, and amber objects and reject the brighter blues, reds, and yellows. A male that places a green chili in his bower even removes it once it ripens to red. Spotted bowerbirds use just about anything of the appropriate size and color: screws and bolts, coins, thimbles, spoons, forks, knives, toothbrushes, car keys, and children's toys. The thieving habits of these birds are well known throughout Australia. One reportedly stole a glass eye from a cup of water and deposited it on its bower! Males also paint their bowers. They prepare a mixture of dried grass and saliva in their mouths, then apply the reddish-brown paste by sliding their bills up and down along the grass stems lining their avenues.

Females make the rounds, inspecting all the neighborhood bowers. Once a female chooses one lucky guy, she returns to his bower doorstep. He sings, jumps up and down, dances erratically, picks up a bone, shell, or other object in his bill, jerks his head up and down, and shakes the object. He flings the object down and then picks up another and does the same.

Meanwhile, the female scrutinizes the male's antics. If she likes what she sees, they mate. After copulation, the female leaves and the male calms down. Tidying up, he returns the displaced objects to their proper places. He's ready to impress another female.

Male satin bowerbirds from Australia are black, with a gloss of violet purple or blue; females are ashy gray-green with a faint tinge of blue. At the beginning of the breeding season, the male satin bowerbird carefully clears a site of leaves, twigs, stones, and any other extraneous material. He then weaves sticks and twigs into two walls about two inches thick and a foot high, forming an avenue about five inches wide. The walls arch inward and nearly cover the avenue. He sets aside one entrance for the display area. This area plus the length of the avenue can extend three feet.

Male satin bowerbirds prefer blue for decorating their bower entrances: flowers and berries, plus bits of blue paper, glass, and ballpoint pen caps. A single bower may have as many as seventy blue parrot feathers. While camping in Lamington National Park in southern Queensland, a friend of mine accidentally left her blue toothbrush outside overnight. The next day Kay found it at the entrance of a satin bowerbird's boudoir in the nearby woods.

Satin bowerbirds' second favorite color seems to be yellow-green: flowers and fruits, straw and wood shavings. If available, males add snake skins, fungi, snail shells, abandoned wasp nests, and bits of aluminum foil. They prefer unusual objects, and the rarer the better. After a male places a trinket on his bower, he steps back and cocks his head to one side as if surveying his work. If not fully satisfied with the effect, he moves the object and steps back to inspect again. Satin bowerbirds paint the inside of their avenue walls by mixing charcoal dust or fruit pulp (blue, of course) and saliva, and then dabbing this paste onto twigs. Unlike spotted bowerbirds that apply paste with their bills, satin bowerbirds use frayed pieces of bark as paintbrushes to apply the paste. They refresh the paint daily.

Male satin bowerbirds arrive at their bowers at dawn and clear away fallen leaves and shift locations of display items. They stay near their bowers most of the day, guarding them from trespassing males that attempt to tear down the walls or steal treasures. (Males of many species of bowerbirds, not just satin bowerbirds, steal trinkets from other males' bowers and tear apart the bowers themselves.) A male leaves his own bower by day only to raid other bowers or to forage for berries and other fruits.

When a female comes to inspect his bower, the male prances around at the entrance. Picking up a blue parrot feather, snail shell, or other bauble in his beak, he arches his tail, makes a whirring noise, and hops about stiffly. His blue eyes bulge and their color intensifies. He launches into a courtship song consisting of buzzes and other mechanical sounds, and then he mimics other birds' songs—including the hoots of kookaburras,  commonly known as laughing jackasses. As the female watches this spectacle from inside the bower and admires his collection of trinkets, her excitement heightens. Eventually the male runs to the rear of the bower. If the female decides this male is the one, she crouches down and waits to copulate. If she votes no, she escapes out the front entrance.

BOWER CONSTRUCTION AND decoration by male bowerbirds might seem excessive, and its daily maintenance compulsive, but not from the birds' perspective. Their bowers are the key to their reproductive success. Stealing treasures and wrecking each other's bowers might appear unnecessarily aggressive but, again, not from the birds' perspective. Females prefer males with the most bric-a-brac on their bowers, thus a male increases his chance of mating by making his neighbors' bowers appear less attractive. Since bowers determine a male's reproductive success, and since males compete so strongly among each other for mates, we shouldn't be surprised that their rivalry leads to thievery and vandalism. After all, these behaviors are exemplified by humans also, an unfortunate consequence of materialism.

SNEAKERS AND DECEIVERS

Encouraged by the tameness of the animal Europa ventured to mount his back, whereupon Jupiter advanced into the sea and swam with her to Crete. You would have thought it was a real bull, so naturally was it wrought, and so natural the water in which it swam.

THOMAS BULFINCH, *Bulfinch's Mythology*

Jupiter, king of the Roman gods and ruler of the universe, was infamous for his sexual appetite. His wife (and sister) Juno was understandably jealous of her husband's extramarital affairs with goddesses and mortals. To deceive her, Jupiter carried out many of his affairs in the guise of a nonhuman animal. He appeared to Europa as a bull, to Leda as a swan, and to Ganymede as an eagle. Fairly clever strategy: alter your appearance and, by Jove, you might get away with it! Jupiter's strategy is not limited to the gods. Biologists call it an "alternative reproductive behavior."

In most animals, males compete among themselves for females. This might involve fighting or defending a prime territory or engaging in elaborate courtship behavior—showing off bright colors, horns, spectacular feathers, or a big body.

So what's a small male to do? Small bull elephant seals and elk sneak into larger males' harems and seize sexual trysts. As we'll see here, small males of many species, some that have harems and some that don't, are adept at sneaking around larger males. Another trick is to mimic a female's appearance or behavior and mate when another male's back is turned. This strategy works even for males of reasonable stature. Consider the following examples of sneakers and female mimickers, just a few from a wide diversity of animals that display alternative reproductive behaviors.

MARINE SOW BUGS (isopods) are closely related to terrestrial pill bugs that you might know as "roly-polies." A friend and colleague at Northern Arizona University, Steve Shuster, studied an unusual species of marine sow bug (*Paracerceis sculpta*) from the Gulf of California that displays both female-mimicking and sneaking behavior. Like roly-polies, these sow bugs conglobate—roll up into balls. Steve's face lights up when he describes these sow bugs to me. "Unlike the drab, gray roly-polies, these guys are gorgeous, like little jewels." They come in a rainbow of colors, including orange, purple, green, red, and iridescent blue. Some have spots, others flashy racing stripes. Steve calls one striped purple-and-white variety the "purple zebras." Others are pure snow-white with jet-black heads.

These colorful sow bugs mate inside so-called breadcrumb sponges that live beneath boulders and ledges in permanent tide pools. The whitish sponges resemble the innards of day-old French bread—they're rather crispy in texture and are loaded with large holes that provide apartment complexes for small invertebrates. A male sow bug mates with as many females as he can, but a female mates only once and broods her young inside the sponge.

Steve discovered that a male sow bug's reproductive behavior varies depending on which of three discrete sizes he is. A large alpha male, the size of a pinto bean, stations himself at a breadcrumb sponge and attracts lentil-sized females. He guards and defends his harem of up to eleven females inside the sponge. Smaller beta males, the same size as females, invade the sponge by mimicking the appearance and courtship behavior of sexually mature females. Since the alpha male mistakes them for females, he rarely ousts the betas. Tiny fast-moving gamma males, the size of sesame seeds, zip into the sponge. Steve calls these guys "swimming testicles" because their tiny bodies are chock-full of sperm. The alpha male chases the gammas if he notices them, but often they hide in crevices. These alternative behaviors are successful, though beta and gamma males have a better chance of fertilizing females in large harems—presumably because (like an elephant seal harem master) the alpha male can't monopolize every female.

Another fascinating marine animal with alternative mating behaviors is the giant cuttlefish, closely related to squid and octopuses. Whereas squid have torpedo-shaped bodies and octopuses are bulbous, cuttlefish are shaped like shields. They have broad heads, large eyes, and eight rather short arms and two long tentacles surrounding the mouth. A shell called cuttlebone grows inside the soft body (the very same cuttlebone given to canaries and parrots as a calcium source). Recent observations made in a rocky reef area off South Australia reveal that male giant Australian cuttlefish sometimes sneak matings or mimic females to achieve matings.

Female giant cuttlefish attach their three-quarter-inch white eggs individually onto the undersides of rocks or in reef crevices. They lay successive eggs about six to twenty-two minutes apart. As they swim around searching for egg-laying sites, the already-mated females mate with additional males. Some females mate four times before laying an individual egg. It's unknown how many eggs a female lays during the breeding season, but marked females laid eggs for up to six weeks. Females have sperm receptacles, presumably allowing for sperm storage from multiple males. (We don't know the consequences in giant cuttlefish, but in some other animals in which the females store sperm from multiple males, females seem to control which sperm fertilize their eggs. This selection occurs within the reproductive tract and may involve chemical, immunological, or physical factors. The female reproductive tract may also serve as an arena where sperm from different males compete among themselves to fertilize the eggs.)

Within the aggregation of spawning cuttlefish studied, males exhibit different behaviors depending on their size. Most large males (more than ten to nearly fifteen inches) pair with females, which they guard both before and after copulation. Although pair formation lasts from seven seconds to ninety minutes, most are from five to ten minutes. During this time, males hover over the females or station themselves between the females and other males. They flare their arms and arch their bodies to protect their mates from other males, and they shove and bite if provoked. Small males (six to ten inches, the same size as females) generally try to sneak copulations, a more successful strategy than pairing with and defending females since large males usually displace small males.

Small males sneak copulations in several ways, and an individual uses any of the techniques depending on the situation at hand. In "open stealth," a male hovers and watches a pair until the paired male is distracted, at which point the sneaker male darts in and overtly attempts to mate with the female. In "hidden stealth," a male hides under a rock and attempts a covert mating with any female that approaches searching for an egg-laying site. In "female mimicry," a male mimics the coloration and posture of an egg-laying female and approaches a pair without being challenged. When the paired male is distracted, the small male tries to mate. Large paired males are so fooled that they defend and even try to mate with these small males. The investigators found that, on average, four out of every five males were sneakers. If they were as successful as the big guys, we would expect sneakers to have achieved four-fifths (80 percent) of the matings observed in the study. They didn't get that many, but they got almost half. So, although they weren't as successful as the big guys, sneakers' techniques obviously work.

Fish also display alternative reproductive behaviors. Have you ever waded in shallow water with spawning bluegill sunfish? If so, you've seen their crater-shaped nests and had males dart at you and nibble on your toes. These are "parental" males, conservative stay-at-home dads that build and defend nests. That's why they nibbled on your toes.

Male bluegills display one of three different mating behaviors. Parental males become reproductively mature at about seven years old and six or seven inches long. At that point, they build nests and court females. After successfully attracting a female to his nest, a parental male fertilizes her eggs and stays a week or so guarding and fanning his thousands of youngsters to ensure they're well oxygenated. The other two types of males, "sneakers" and "satellites," mature by about their second birthday when they reach almost three and four inches, respectively. Males of both types parasitize parental males' nests. Sneaker males cuckold parental males by lying in wait behind plants or rocks, darting into nests while the females are spawning, and attempting to fertilize some of the eggs. Satellite males cuckold parental males by resembling females in color and soliciting parental males for courtship. While in the nests, they also act like males and court the females. This double courtship often allows the satellites to fertilize some of the females' eggs while not arousing the parental males' suspicions. There's a genetic basis to the alternative mating behaviors: cuckolder males are more likely than parental males to produce sons that take up cuckoldry.

A male South American leaf fish can mimic a female by changing his behavior and color, and, unlike the bluegills, leaf fish can change strategies. At times a male can be a normal territorial fellow and at other times a pseudo-female. A territorial male is nearly black with pearly white spots. In contrast, a reproductively ready female is yellowish pink to white with scattered small brown spots.

Once reproductively mature, a male leaf fish defends his territory—an area suitable for spawning, such as a crevice between rocks. He lies in wait just inside his lair, watching for intruding males to fend off or females with which to mate. After entering a male's lair, a receptive female turns upside down and places her back end against the underside of a leaf or rock and releases eggs. The male stays underneath her and discharges sperm up onto the eggs. Additional females may approach the spawning couple. The territorial male usually chases them away, but occasionally a female slips through and the male spawns with both females simultaneously—ménage à trois fish style.

But then there are males that—like Jupiter, beta sow bugs, small cuttlefish, and satellite bluegills—disguise themselves in order to mate. Male leaf fish sometimes mimic female coloration and behavior and approach spawning couples. When this happens, the territorial male swims toward the intruder, but he's less aggressive than he would be to a typically colored dark male. The intruder acts like a female and doesn't return the hostile displays. If lucky, the pseudo-female enters the lair and squirts some of his sperm up onto the eggs.

Some amphibians engage in deception and sneaky sex. The final stage of courtship in tiger salamanders is a "tail-nudging walk," during which the female follows the male along the pond bottom and nudges his tail. Because the male can't see the female, he relies on her nudges to keep track of her. Once the male determines that the female is receptive, he deposits a spermatophore (packet of sperm) on the mud and moves forward. The female walks over it and draws the packet into her body.

Sometimes male tiger salamanders mimic females and trick other males into depositing (and wasting) their spermatophores. A second male moves in between a courting pair and imitates female behavior by nudging the male. The lead male, thinking the nudger is the female who was right behind him moments earlier, deposits a spermatophore. The intruding male then deposits his own sperm packet on top of the first. The female, who is now nudging the intruder, picks up the top spermatophore, leaving behind the original one from the duped male.

Bullfrogs, some toads, and some treefrogs exhibit satellite behavior: males sit near dominant or territorial males that are calling to attract females. In some species, the lurking male is simply waiting for a good calling site. As soon as the resident male attracts a female, the satellite male begins calling. In other species, the satellite male waits for the chance to intercept females without the energetic expense of calling. In some species, a satellite male is a small individual. This may be his best chance to breed until he grows larger. Although he may not have a very good chance of mating, a small chance is better than none. In other species, a male may switch strategies. For a few nights he might actively call to attract a mate, then the following night he might be a satellite. In this case, satellite behavior may conserve energy yet give a male a chance to mate.

The spring peeper is a small brown treefrog common in the eastern United States. Large choruses of calling males form in water-filled roadside ditches and ephemeral ponds, creating a cacophony of high-pitched peeps. If you go out on a rainy spring evening and watch these frogs closely, you're likely to see certain males hunkered down, in a crouched posture, within eight inches of actively calling males. The more frogs at the site, the more satellites you'll see, so be sure to go out on a night spring peepers love—the wetter and more miserable for you the better. Are spring peeper satellite males physiologically inferior to calling males? This question is the title of an article published in 1993. The answer? The investigators found no evidence that satellites are inferior. Callers and satellites were not significantly different in body size or in physical condition as measured by muscle mass and muscle enzyme activity. Spring peeper satellites aren't wimps. They may simply save energy by duping other males for the night.

AND SO IT is with mortal animals—sow bugs, cuttlefish, fish, salamanders, and frogs—that sneak in and seize a mating opportunity or mimic females in clever deception. Their strategy is not unlike Jupiter's. Sometimes sneaky sex is the best way to go.

If one mating behavior results in greater reproductive success, then why don't alternative behaviors disappear over evolutionary time? Why don't all individuals exhibit the best strategy? Because environmental conditions and social interactions among individuals change constantly, the behavior with the best payoff may depend on what other individuals are doing and how many individuals are using this same approach. For many species, it may be advantageous to have several options so that individuals can "play the odds." Nonsneaky individuals persist in a population because sneaky sex isn't the best way to go for all individuals all the time.

HEADLESS MALES MAKE GREAT LOVERS

The Praying Mantis

From whence arrived the praying mantis?
From outer space, or lost Atlantis?
I glimpse the grim, green metal mug
That masks this pseudo-saintly bug,
Orthopterous, also carnivorous,
And faintly whisper, Lord deliver us.

OGDEN NASH, from *Custard and Company*

Continue having sex after your head has been chewed off? That's what intrigued my fifteen-year-old son. He was writing a report on the sex lives of praying mantids—the first time all semester I'd seen him do any biology homework.

Rob knew that mantids are relentless predators. As a kid, he'd seen them lie in wait for prey, with their spiny, grasping front legs upraised and their pincers partially opened, poised for action. He'd watched mantids' heads swivel three hundred degrees as the huge eyes tracked their moving prey. And he'd witnessed unwary grasshoppers and caterpillars get nailed and then consumed. He'd read that female mantids sometimes eat their mates, but until freshman biology class he hadn't known that headless males make great lovers.

Imagine the following scenario: a male mantid, attracted by the smell of a female hidden in a lilac bush, creeps up behind her and when close enough leaps onto her, secures a perfect grip on her body, and copulates.

No courtship, no permission asked or granted. He has behaved "appropriately" for a male mantid.

If a male doesn't behave appropriately, he may incite trouble. Positioning is everything. A male that approaches a female from the front may meet immediate death by decapitation. If he sneaks up behind her but is just a little off on his grip, the female might bite off his head and dine on her brainless suitor as he continues to pass sperm into her body. Sometimes the impetuous female partially eats the male before he even mounts her. In this case, the headless wonder swings his legs around until his body touches hers, climbs onto her back, and copulates as though nothing were amiss.

Headless sex? How can it be? Copulatory movements in mantids are controlled by masses of nerve tissue in the abdomen rather than the brain. Males of some mantid species mate more effectively when decapitated. Why? A nerve center in the male's head inhibits mating until a female is clasped. If this nerve is removed, such as when the female bites off the male's head, all control is lost and the result is repeated copulation.

Sometimes the female devours her mate under circumstances outside the male's control. If the pair is disturbed and the temperamental female becomes frightened, her immediate reaction is to whip around, snatch the male's head in her greedy mandibles, and gnaw it off. In some species, a female's propensity to consume her mate is unrelated to the male's behavior or outside disturbance. It's simply part of the mating ritual.

In all fairness to the eighteen hundred or so species of mantids, cannibalism is far from universal. Sexual cannibalism—defined as a female killing and eating her mate during courtship, copulation, or shortly after copulation—probably occurs in a minority of mantids.

Considering mantids' potential cannibalistic behavior, why do people worldwide endow these insects with holiness? It's their characteristic posture of supplication—front legs held together outstretched toward heaven—that commands respect. People living in the Middle Ages believed that mantids spent most of their lives praying to God, and some Muslims insisted that mantids always pray with their heads facing Mecca. *Mantis* is a Greek word meaning "prophet" or "clairvoyant." Soothsayers used mantids, believed capable of foretelling events, to distinguish the fortunate from the unfortunate. Ancient Egyptians believed that mantids carried a person to the gods after death. Some Africans worshipped mantids because they believed these insects could resurrect the dead. Some ancient beliefs are still with us: If a mantid alights on your hand, you'll meet a distinguished person; if it alights on your head, you'll receive a great honor.

Even the mantids' cannibalistic reputation has earned them respect. The Asmats, a native tribe that lives in southern New Guinea, greatly admire these insects. Until the mid-1940s, Asmats were fierce headhunters and cannibals, so it was natural they should admire an insect as fearless and cannibalistic as themselves. To honor these insects, Asmats commonly decorated their drums, shields, and spears with mantid figures.

ROB'S QUESTION WAS *why* do some females eat their lovers? Biologists, long fascinated by sexual cannibalism, have suggested various explanations for why such bizarre behavior should evolve and be maintained in some mantids, spiders, and other invertebrates. There are different, nonmutually exclusive explanations depending on the species and its natural history and whether the female eats the male before or after mating.

Mistaken identity may explain why females eat their prospective mates before copulation. Females that engage in sexual cannibalism are aggressive and predaceous by nature, and they may mistake a potential mate for food. This rationalization probably fails to explain many instances of sexual cannibalism, however. If females couldn't discriminate between potential mates and food, many would end up virgins and never reproduce.

An alternative explanation for cannibalizing a prospective mate before copulation is discrimination among potential mates. Female garden spiders are more likely to cannibalize small wimpy males than large macho males. Why doesn't a female just reject the small guys and allow them to move on? The small ones make nutritious treats. Just one male, even a small one, gives the female energy to produce more eggs.

A third explanation for why a female might eat her suitor before copulation is that perhaps she weighs the value of a courting male as a sperm donor versus a food item. If she hasn't been very successful finding food recently and it's early in the breeding season, she should eat the male. On the other hand, if courting males seem to be a rare commodity and especially if it's late in the breeding season, she should mate with him. Females that have already mated once would be expected to attack courting males more often than should virgin females. A recent study with fishing spiders lends some support for this idea but does not exclude the possibility that sexual cannibalism may be simply misplaced aggression.

Why would a female eat her mate after copulation? In some cases, males might offer themselves up for food as an investment in their offspring, especially if they have little chance of mating again. For example, female orb-weaving spiders often eat their mates after copulation, but even if a male is not attacked, he will die soon after his first mating. A male might increase his reproductive success if the extra nutrition provided by his body increases the number of young his mate can produce or if it improves his offspring's health, size, or survival. This explanation at one time was suggested for black widow spiders. They've been falsely accused, however. Sexual cannibalism is not as common in these spiders as originally believed. Furthermore, like mantids, male black widow spiders do not willingly sacrifice themselves. Although studies of some species of spiders and mantids have shown a positive effect of sexual cannibalism on the female's reproduction, other studies have shown no effect. Obviously, this is not the rationale for all species.

Another explanation for eating a male during or after copulation is assurance of paternity. In many biting midges (tiny flies called no-see-ums or sand flies), the female eats the male during copulation, treating her mate the way she does any other insect prey. She pierces his cuticle and dissolves and sucks out his body contents, draining him in about thirty minutes. When the female disengages from her lifeless mate, a portion of his body remains attached to her. This "plug" may keep other males from mating with her, thus assuring not only paternity of the now-deceased male but also fertilization of all the female's eggs. By dying in this way, the male increases his reproductive success.

If a male's fitness benefits from his being eaten during or after copulation, he might facilitate cannibalism, especially if he is unlikely to mate again. One example where males actually encourage sexual cannibalism is the Australian redback spider (*Latrodectus hasselti*), closely related to the black widow spider.

Mating in Australian redback spiders occurs on the female's web. After a male probes, taps, and nuzzles the female, he inserts a palp (the appendage that transfers sperm) into his much larger mate. Then he flails his legs in the air and turns a somersault, bringing his abdomen up against the female's mouthparts. He effectively says, "Here I am—eat me." The female slowly consumes her mate during copulation. The male withdraws once he has completed sperm transfer five to thirty minutes later. Mutilated, he grooms himself in preparation for the next lovemaking session. About ten minutes later, he briefly probes, taps, and nuzzles his mate and then engages his second palp. Again he somersaults and presents his now-shrunken abdomen to his mate. She sinks her mouthparts into his body and continues to digest. After completing insemination, the now-weakened male withdraws, and the female wraps him in silk. It will take her less than fifteen minutes to finish her meal.

A field study of Australian redback spiders reveals that females eat their mates about 65 percent of the time. Males always do the somersault, offering themselves up for a meal, but only hungry females accept the offer. A male is only 1 to 2 percent of the female's mass, and cannibalism by the female does not increase the number or mass of eggs she produces. So why such behavior from the males? Cannibalized male spiders receive two paternity advantages. First, they copulate twice as long and therefore fertilize about twice as many eggs as uneaten males. Second, cannibalized males are more likely to father offspring because females are less inclined to remate after eating a male. If a female mated again, the second male would father some of her eggs. The cost of suicide is low for these spiders because even if they survive a copulation, they're unlikely to mate again. Whereas females live up to two years, males live only two to four months after they mature. Males rarely eat after reaching maturity, and if they survive a mating, they often stay on their mate's web.

AFTER MY SON turned in his report, I enlightened him further on the macabre behavior of sexual cannibalism. "Real guys don't lose their heads," I confided. "They have ways of avoiding being eaten by their mates."

Some scorpions use a variation on the "wham, bam, thank you ma'am" technique. Female scorpions release sexual attractants that guide males to them. Once a male locates a female, he grasps her claws with his and leads her in a prolonged mating waltz, referred to as a *promenade à deux*, in search of a hard surface, a stick or rock. Once the scorpions find an appropriate object, the male deposits a sperm-bearing packet on the object and then drags the female over it. She sucks the sperm into her genital pore, and almost immediately the male violently smacks her with his tail and scurries off. Why? Because if he isn't fast enough, she may eat him.

Male spiders have trouble even getting close enough to a female for sex. The minute a female feels vibrations on her web, she assumes food has arrived. Males are almost always smaller than females, so they're often mistaken for food. Males must identify themselves as males before the females' hunting instincts and gastric juices take control.

In some species, the male vibrates, twitches, or drums on the female's web in a characteristic way that announces he's a potential suitor rather than food. Sometimes a male wolf spider must signal his intentions by waving his legs for hours, circling a female at a discreet distance, before she submits. If a male stumbles upon a female unexpectedly and has no time to wave his legs, woe unto him! Male jumping spiders signal their intentions to females through complex courtship dances. They zigzag to and fro, jerking their abdomens about in styles reminiscent of Hawaiian hula dances. Some just sway as if in a drunken stupor.

Some male spiders don't announce their presence at all but have ingenious ways of avoiding being cannibalized. For some, patience pays off. A male waits on the sidelines of the web until the object of his desire has captured something to eat. Then, while she's otherwise occupied, he makes his move.

Bribes also do the job. Some spiders offer nuptial gifts to their prospective mates. Once a male common European nursery web spider matures, instead of eating a juicy insect, he wraps it in silk, holds the enshrouded corpse in his jaws, and searches for a female. Once she's found, he approaches her cautiously and offers his gift. If the offering is too small, the female walks away. If it's acceptably large, she bites into it and begins to suck out the juices. Once she's distracted by eating, the male mates with her.

A little brute force also works wonders. Courtship in some tarantulas begins with the male drumming his front legs against the female's body. Though she lifts her front legs in a threatening gesture, the suitor taps and strokes her. She responds to his overtures by lifting her body higher and exposing her fangs. A stab by the female could prove fatal, so the male wraps a pair of hooks on his front legs over her fangs, yielding them ineffective. In this position, he can safely transfer sperm.

Would you believe kidnapping as a strategy? Males of one crab spider kidnap immature females and build silken tents around them that serve as prisons. Standing guard over their captives, these males chase away other males. Once the females are barely mature, the males assault them before they're strong enough to resist.

And then there's good old-fashioned trickery. Males of some spiders, including crab spiders, slowly creep up on females and throw a silk net over them before mating. After mating, the male escapes while the impregnated female disentangles her legs. These males can depart nonchalantly since the females are roped down. In most other spiders, the male dashes seemingly panic-stricken from his ex-lover. If he doesn't escape quickly, she may tie him up, kill, and devour him.

The prize for cleverness and ingenuity in avoiding sexual cannibalism goes to male dance flies. These common small to medium-sized flies often swarm around trees and fly in an up-and-down or circular motion, thus the common name. Females are hotheaded and aggressive—a bad combination for male dance flies. As in some spiders, distraction is the solution. And males have developed brilliant tricks.

Most dance flies eat other small flies, such as mosquitoes and midges. In some species, the male dance fly offers the female a nuptial gift—an insect he has caught. He can then mate safely while the female is distracted by eating. In other species, the male is more imaginative and artistic. He wraps his gift in silk before offering it to the female, gaining precious time as she unwraps the insect before consuming it. We've all known stingy people. But dance flies? One species captures an insect, sucks out the juices for his own meal, and then wraps it in silk. He offers a female the parcel and begins to copulate. By the time she has unwrapped the empty insect shell, he has completed mating and even gotten a meal to boot.

I THINK I convinced my fifteen-year old son that the mating game can be dangerous. As he walked away, I imagined him thinking, "What if you gave a girl a Whitman's sampler box with nothing left but the papers? Or sucked out the cream insides and gave her the chocolate shells?"

TRADING FOOD FOR SEX

The giving of chocolates carries with it an assumption of expectation of **intimacy**. It is to melt in the mouth of the recipient, and intoxicate the senses. Chocolate, in both essence and mythology, is an aphrodisiac. It is a fond declaration of love, and a plea for its reciprocation.

CHRIS SAVAGE KING, quoted in Linda K. Fuller,
Chocolate Fads, Folklore, & Fantasies

We've just seen that some male spiders and dance flies offer food to their prospective mates as a distraction. The male avoids being cannibalized by mating while the female eats. What about chocolates given by a human suitor? Presumably such an offering is not to avoid sexual cannibalism, but a cynic might suggest that the gesture is more than simply symbolic.

Certain insects and birds offer nuptial gifts to prospective mates—not to avoid being eaten, but as bribes. Nuptial gifts may be bits of nesting material or random objects, but generally the way to a female's heart is through her stomach. Nuptial food gifts may be crucial to successful breeding. The quality of the gift may translate directly into a nutritional benefit for the female—the larger the gift, the more eggs she can produce. This is the case for an Australian bushcricket (*Requena verticalis*).

The male bushcricket transfers an elaborate spermatophore (packet of sperm) to the female. The spermatophore consists of two parts: an ampulla that houses the sperm, and the spermatophylax, a gelatinous sperm-free mass rich in protein. Immediately after the male attaches the spermatophore to the base of the female's ovipositor, she sticks her head between her legs and begins to eat the spermatophylax. While she eats, insemination occurs. The larger the spermatophylax, the longer it takes her to eat it (just as it takes you longer to eat a twelve-ounce chocolate bar than a Hershey's Kiss). The more time she spends eating the spermatophylax, the more sperm enter her sperm-storage organs. Thus, the bigger the nuptial gift a male offers, the more eggs he fertilizes. On average, a female spends about fifty-two minutes eating the spermatophylax. Once she's finished, she eats what's left of the ampulla. For this reason, the spermatophylax also functions to prevent the female from eating the sperm-containing ampulla prematurely.

Scientists in western Australia have discovered that these bushcrickets produce spermatophylaxes of different sizes. When females are abundant, males offer gifts of a size that, on average, is just large enough to feed females to achieve complete sperm transfer. When females are scarce, males offer gifts twice the size necessary to ensure insemination. Male bushcrickets have different ways of maximizing their reproductive success depending on availability of females. When their chance of multiple mating is high, males invest just enough in the gift to allow for sperm transfer. That way they can mate more often. When the chance of multiple mating is low, males invest more in their offspring. Larger spermatophylaxes have a higher concentration of protein and are more nutritious than smaller ones. The larger the spermatophylax a female eats, the more offspring she will produce.

Hangingflies also offer gourmet meals in return for sex. Gift giving in these insects entices females to copulate and ensures sperm transfer. A male's gift is unlikely to benefit the eggs he fertilizes, because the female lays eggs within three to four hours after mating.

The hangingfly *Hylobittacus apicalis* lives in deciduous forests throughout much of the eastern United States. When they're not flying, these nearly one-inch-long insects hang from a leaf or twig by their long forelegs, thus the common name. (Hangingflies are not true flies. Flies have a single pair of wings; hangingflies have two pairs. True flies belong to the order Diptera; hangingflies belong to the order Mecoptera.) Hangingflies eat mainly other insects. They inject digestive juices into their victims and suck out the goodies through their long proboscises. Female hangingflies hunt for their own food during the nonbreeding season, but during the mating period, they rely heavily on males to bring them nuptial gifts.

When a male hangingfly catches a small insect, he devours it. When he catches a large insect, he only samples it. After sampling a large insect, he carries it with his hind legs and flies to a plant, where he hangs from his front legs and releases a sex attractant from his abdominal glands. He waits. If no female arrives, he flies a short distance, still carrying his gift, and releases more attractant. If he's lucky, a female eventually alights on the plant and hangs facing him. Romeo offers his gift.

What happens next depends on the size and quality of the gift. If the gift insect is palatable and at least nearly the size of a housefly, the female allows her suitor to copulate for twenty minutes or longer. Once finished, the male disengages his genitalia and then wrests what's left of the insect from her grip. If enough is left over, he'll offer it to another female. If only a little is left, he'll finish it himself. What if the gift is puny? Or what if it's a ladybug or some other nasty-tasting insect? A male gets what he pays for. In those cases, the female pulls back her abdomen and refuses to copulate or will do so for only a few minutes. Five full minutes of copulation are required for sperm transfer, so the female effectively refuses sperm from this male. By preferring males that offer large good-tasting nuptial gifts, a female probably lives longer because she doesn't endanger herself by hunting for food as often as she would otherwise.

This is the "normal" behavior, but some male hangingflies take the easy route and turn to piracy: they mimic female behavior (easily done, as males and females look alike) and steal insects from rival males. A pirate approaches a male and lowers his wings, the cue for the male to offer his gift. The male responds, and the pirate steals the insect. This sneaky behavior is quite productive. Because they don't have to hunt for nuptial prey, female-mimicking males have time on their wings. Also, since they're not flying through the trees looking for nuptial gifts, they're less likely to get caught in spiderwebs. They may live longer than their more honest counterparts. So why don't all male hangingflies turn to piracy? There's a downside. It often takes a long time to locate a gift-bearing male, and once one is found, the sucker must be conned. The pirate doesn't always get away with his crime.

Scorpionflies are close relatives of hangingflies. The male's genitalia usually curves upward and forward, resembling the sting of a scorpion. These insects feed primarily on dead insects, though they also eat fruits and nectar. Like hangingflies, male scorpionflies generally offer food to their prospective mates. These gifts probably function in the same way as in hangingflies: to entice the female to mate and to maximize sperm transfer. Copulation lasts longer if males offer gifts. Longer copulation means that more sperm are transferred and more eggs are fertilized.

Different species of scorpionflies offer different nuptial gifts. Some males give dead insects. The best place to find a dead insect is in a spider's web, but stealing the corpse can be risky. Males of other species eat dead insects and then produce nutritional saliva in their large salivary glands. At the appropriate moment, they offer masses of saliva to their prospective lovers. Copulation lasts longer if a male offers more than one saliva mass. In some species, a male offers a dead insect if he can find one, or if not, he offers saliva.

Males of some North American and Japanese scorpionflies display an alternate mating behavior. If a male can't find a dead insect and if he is unable to secrete a mound of saliva, he takes a female by force. He rushes at a female, grabs a leg or a wing, repositions and pins her down, and then copulates. Since captured insects and saliva masses cost something to a male, why don't males always use the "rape" technique? Recall that copulation is longer when gifts are given, thus more sperm are transferred. A male increases his reproductive success by offering a gift. But if he can't, he does the best he can.

Males of the scorpionfly *Panorpa cognata* attract females by releasing long-range chemical attractants. Once a female flies in, the male wiggles his abdomen, flutters his wings, and sporadically touches his privates to the female's genital area. When the male is ready to copulate minutes to nearly seven hours later, he secretes saliva from his salivary gland. He glues this mass onto a leaf and then guides the female to it. The female eats her free meal while the two mate. But once finished eating, she abruptly stops copulation.

Experiments carried out in southwestern Germany on *Panorpa cognata* reveal that the difference in duration of premating behavior, from minutes to hours, relates to the male's nutritional condition. Males fed on a high-nutrient diet developed larger salivary glands, and these males initiated courtship sooner than males fed on a low-nutrient diet. Why should a male with a small salivary gland prolong the premating period before offering his gift of spit? By delaying copulation, a male is more likely to get a successful

mating in return for his investment. When any male offers his gift right off
the bat, his efforts are likely to be thwarted—either because the female rejects him or because other scorpionflies interrupt the session. A male with
limited resources (a guy not in tiptop condition) invests his gift wisely. He
waits until he is certain that the female is motivated to mate and that intruders aren't nearby. In contrast, each saliva mass is less valuable to a
male in good condition. His strategy is to rush, in the interest of mating
with more females. If he loses his spit gift, he can easily produce another.

LIKE INSECTS, BIRDS are intriguing nuptial gift givers. Males of
many species of land birds and seabirds feed their prospective mates during courtship.

Consider the common tern, an abundant seabird in North America
that nests in colonies. These terns are mostly white and gray, with long
narrow wings tipped with black, a forked tail, a black cap, and a pointed
bright red-orange bill. They're monogamous, with a "divorce rate" of
about 20 percent from one year to the next.

Common terns engage in three phases of courtship. The first is pair
formation. During this phase, the male parades up and down the beach
with a fish crossways in his bill, displaying to any unattached female who
will watch. If a female appears interested, at least in the fish, he attempts
to lure her to his territory. If successful, he continues to display with his

fish. Initially he's reluctant to
give up his prize, but as the
pair-bond strengthens, he
offers the fish. After she eats
it, the pair copulates.

The second phase is the honeymoon, lasting five to ten days. During
this phase, the birds spend much of the daytime on the feeding grounds.
The female stands on the beach or on a rock and waits for her mate to deliver fish or shrimp to her. She rarely catches her own food during this
phase. Because a female must have gathered most of her weight reserves
by the time she lays her eggs, courtship feeding provides a substantial
contribution to her nutritional state.

The third phase is egg laying, during which the female usually stays in
the nesting territory, distant from the feeding grounds. After one to six
days, she lays her first egg, then another egg two days later. Some females
lay a third egg one or two days after the second. The male is attentive to
his mate during this phase and frequently brings her fish or shrimp. Because the two are not together, the male expends more energy feeding his

mate during this third phase than during the honeymoon period. He must commute to the feeding grounds, catch a fish or shrimp, and then fly back to the nesting area with the food—an avian pizza delivery boy.

The number of eggs a female lays depends largely on how much food she eats during the honeymoon phase. Given that the more food she eats, the more eggs she lays, why doesn't a female supplement her diet beyond what the male provides? She may be handicapped by her weight: a female that lays three eggs weighs almost 50 percent more than her normal weight. Catching a fish is no easy trick for a tern. It must cruise over the water at low speeds and precisely control its dive. The effort required to attack a moving target might be just too much for an egg-laden tern.

Courtship feeding in common terns has three functions. It helps to strengthen and maintain the pair-bond. It also contributes significantly to the nutritional reserves of the female during egg formation. In addition, it may allow the female to decide whether or not to stay with a particular male. Once the eggs hatch, the female broods the chicks and the male feeds them. Because a male's performance during courtship feeding is a good predictor of his ability to feed his chicks, a female may judge a potential mate's future ability as a father by his performance during courtship feeding. If he seems incompetent, she might bolt and find a better provider. She might use the second phase of courtship to test the male's fishing ability and the third phase to test his dependability in bringing food back to the nest.

Red-billed gulls, from Australia and New Zealand, are another of the many seabirds that offer courtship feeding. The female begs by hunching over, walking toward her mate, and bending her head down until her bill nearly touches the ground. In a dramatic display, she tosses her head back, opens her bill, and squawks. The male at first ignores the request. The female then makes short pulling movements with her bill at the corner of her mate's bill. No longer able to ignore his demanding mate, the male opens his bill wide and regurgitates a meal of little shrimplike crustaceans. She reaches in and helps herself. The male feeds his mate up to four times per day. Each time he regurgitates up to an ounce of crustaceans, representing a substantial amount of the female's daily caloric intake.

Courtship feeding in these gulls has a second function. Because a female is more receptive after a feeding, a male trades food for sex. The more food regurgitated during a feeding bout, the more likely the copulation will be successful. A male's performance is critical for his reproductive future. Well-fed females return to their partners the following breeding season; females that are not well fed find new mates.

As a final example of courtship feeding, let's look at ospreys. These brown-and-white fish-eating birds live near lakes, rivers, and seacoasts throughout much of the world. Ospreys are large, up to two feet long, with a wingspan of almost six feet. Normally monogamous, these birds have unusually high rates of copulation to produce one clutch of three eggs. They also have intense courtship feeding, where males deliver fish to their partners back at the nest. A study carried out in Corsica, in the western Mediterranean, reveals an association between these two behaviors. Ospreys trade food for sex, and this behavior appears to encourage a monogamous relationship.

In the Corsica population, pairs copulated over a period of forty-five days, and they averaged 288 copulations for each clutch of eggs. Females solicited both feedings and copulations with loud calls. The more fish a male brought and the larger the fish, the more copulations occurred. Copulations lasted longer and were more likely to be successful after a courtship feeding than after a male's return to the nest without a fish delivery. Copulations lasted longer if a large fish were offered as compared to a small fish. As with common terns, a male's performance during courtship feeding reliably predicted how good a provider he was for his chicks.

AS WE'VE SEEN, there is indeed reason behind the apparent madness of males giving sperm-free gelatinous masses, insects, mounds of nutritious spit, fish, and regurgitated crustaceans to their prospective mates. But what about humans and those chocolates?

In 1519 the Spanish explorer Hernando Cortés visited the Aztec emperor of Mexico, Montezuma II. There Cortés learned about chocolate, made from the seeds of the cacao tree, *Theobroma cacao* (Greek words meaning "food of the gods"). (The Olmec of Mesoamerica may have eaten cacao during the first and part of the second millennia before Christ. They may have given it to the Maya, and in turn the Maya gave it to the Aztecs. Chocolate has been appreciated a long time!) The Aztecs drank *chocolatl*, a drink made from cacao seeds ground into a paste, flavored with vanilla and other spices. Reportedly, Montezuma always drank several golden goblets full of *chocolatl* before entering his harem. Hence the belief that chocolate is an aphrodisiac and why it is still associated with love.

Although the ancient Aztecs exaggerated the powers of chocolate (it isn't an aphrodisiac), chocolate contains phenylethylamine (PEA), a natural mood-altering chemical also found in the human brain. PEA is released from the brain when we're happy. Its release causes a rise in blood pressure and increased heart rate, and the effect can range from a mild feeling of well-being to the intoxicating euphoria we associate with love. Eating a lot of chocolate has a similar effect on your body. Women are especially sensitive to the PEA from chocolate, and some chocoholics experience a euphoria not unlike falling in love. Eating chocolate makes you feel happy, which may explain why many people crave chocolate when they're lonely.

Casanova, that legendary Italian lover, often used chocolate instead of champagne as a seduction bribe. A nuptial gift worthy of the gods.

ANY PARTNER WILL DO

To a Lonely Hermaphrodite

Know
Thyself.

JOHN M. BURNS, *BioGraffiti*

According to Greek mythology, Aphrodite (goddess of desire) had a one-night fling with Hermes (god of fertility) in payment for a favor. Their union resulted in Hermaphroditus, a double-sexed being. Hermaphroditus wasn't born double-sexed, however. Here's how it happened.

To conceal her son's birth from Hermes, Aphrodite entrusted Hermaphroditus to the nymphs of Mount Ida, who raised him in a cave. At age fifteen, Hermaphroditus left his forest home to explore unknown lands. One day he came upon a gleaming pool of water ruled by the nymph Salmacis. When Salmacis saw the handsome lad admiring her pool, she was immediately enamored and tried to seduce him. Ignorant of love, he rejected her. Once Hermaphroditus thought he was alone, he stripped and dove into the water. Hiding in nearby bushes, Salmacis was smitten with desire for his naked body. She flung off her clothes and threw herself into the pool. Still Hermaphroditus denied the nymph, though she stole kisses and fondled him. She coiled her body around his like a serpent and pleaded with the gods that she never be separated from this lovely creature. The gods answered the nymph's prayer as the two bodies merged into one with a single face and form: a combination of male and female.

And thus from Hermaphroditus we get the word "hermaphrodite": an organism that has both male and female reproductive organs and produces both egg and sperm cells.

HERMAPHRODITES COME IN two forms. A simultaneous hermaphrodite functions as both male and female at the same time. A sequential hermaphrodite is one sex during an early phase of its life and the opposite sex during a later phase. Both types have some fairly bizarre mating behaviors.

Earthworms are perhaps the best known simultaneous hermaphrodites. Unable to fertilize their own eggs, earthworms must find willing partners. Two earthworms become attracted to each other, not by visual or sound signals but rather through glandular secretions. They position themselves with their undersides touching. Have you ever looked closely at an earthworm? One of its few noticeable body parts is a swollen glandular area called the clitellum (Latin word for "saddle"). This area plays an important role in earthworm sex—it secretes sticky mucus that holds the partners together for the two to four hours while they mate. With their eyeless and earless heads facing opposite directions, they secrete copious amounts of slime. Eventually their muscles contract and droplets of seminal fluid travel from each male pore to the other's clitellar region.

Planarian flatworms are slightly more engaging than earthworms. Planarians are those little flatworms we all studied in high school biology class. They're elongated and flat, with triangular or rounded heads; most are one-quarter of an inch to one and a half inches long. My favorite experiment in tenth-grade biology was cutting a planarian in half and watching it grow into two individuals—regeneration at its finest.

Recently I came across a scientific paper with the intriguing title "Sperm Trading in a Hermaphroditic Flatworm: Reluctant Fathers and Sexy Mothers." What, I wondered, are these flatworms doing? The authors described their study animal (*Schmidtea polychroa*) as a "promiscuous, frequently mating, simultaneously hermaphroditic" creature.

Copulation in these little planarian sex fiends lasts one to two hours, and their behavior varies depending on their recent social interactions. Planarians that have recently been around other planarians generally exchange ejaculates; each gives and receives. A flatworm digests most of the sperm it receives from another individual, so the explanation for sperm trading seems to be that an individual insists on receiving nutritional compensation for the sperm it has donated. Hence the "reluctant father" label.

In contrast, planarians that haven't recently been around other planarians generally don't demand to receive sperm. They're just eager to donate their sperm, and they usually give a large ejaculate. These isolated individuals are also more attractive as females. Why? Because they're less likely to have some other planarian's sperm stored in their bodies, their partners have a better chance of fathering the eggs. Hence the "sexy mother" label. When a planarian is more attractive as a female, its partner donates sperm more eagerly.

Mollusks (snails, slugs, oysters, sea slugs, and their relatives) display unusual mating behavior. Most land snails are simultaneous hermaphrodites, and most must cross-fertilize. Many convey their romantic intentions in peculiar ways. For example, a common land snail (*Helix*) searching for a mate probes its sensory tentacles into nooks and crannies. When it finds another snail of the same species, it gently brushes its tentacles against the prospective lover. If the other snail is similarly inclined, the two raise up, push their large muscular feet together, and stroke each other with their tentacles. Suddenly one snail jabs its partner with a calcareous or chitinous spearlike dart from its vagina. The snail shot by the "love dart" (Cupid's arrow?) retreats into its shell. If it pokes its head back out, the other snail knows that things are lookin' good. The stabbed snail reciprocates by shooting a dart. The snails play footsies again and shoot more darts into each other. Eventually each inserts its penis into the other's vagina.

Land slugs (naked cousins of the clothed snails) are simultaneous hermaphrodites, and many come well-endowed. The uncoiled penis of a five- to eight-inch slug may be more than thirty inches long. Prior to mating, two slugs often fight and display their sex organs to each other. As expressed by slug watchers C. David Rollo and William G. Wellington from British Columbia: "The sight of a courting pair of slugs majestically circling one another and ceremoniously rasping small patches of skin from each other's flanks while they solemnly wave their enormous penises overhead puts the most improbably athletic couples of Pompeii and Khajraho into a more appropriate and severely diminished perspective."

Some slugs pay a price for endowment. Giant slugs (*Ariolimax*) practice *apophallation:* the gnawing off of the penis as a part of mating. When two giant slugs meet, they try to gnaw off each other's penis. If successful, one forces the other individual to become an obligate female because the penis will not regenerate. The individual that gnaws the most penises off other slugs without losing its own increases its reproductive success relative to other individuals. Why? Because it can still function as a male (fertilize other slugs' eggs) as well as a female (produce its own eggs), thus increasing its opportunities to pass on its genes. A giant slug doesn't normally crawl around dangling its organ around behind to tempt fate. The penis is a highly muscular organ kept inside the body and everted only prior to copulation.

Sea slugs are marine mollusks with small internal shells. Most are a few inches to ten inches long. Some are brown or gray and are perfectly camouflaged in their environment. Others are brilliant yellow, orange, red, green, blue, and purple, lavishly decorated with contrasting stripes, polka dots, or splotches.

Many sea slugs mate reciprocally. The partners get into a head-to-tail position and face opposite directions. Each inserts its penis into its partner, and they mutually exchange sperm. In contrast, in some species, only one partner is fertilized at a time, with both individuals facing the same direction. One animal, acting as the female, attaches to a rock. The second animal, acting as the male, crawls onto the first and transfers sperm. Once finished, they reverse roles so both get their eggs fertilized.

Mating in the algae-eating sea slug genus *Aplysia* can get a bit kinky, as other individuals often join the original pair, forming a chain of up to ten mating sea slugs. In a chain, the first animal acts only as a female. Each succeeding animal except the last acts as a male (sperm donor) to the sea slug in front of it and as a female (sperm recipient) to the animal behind it. The last acts only as a male, unless it connects with the first sea slug, forming a ring. In that case, what used to be the first and last animals now function as both sexes. Copulation among chains and rings of these sea slugs can last several days, and the fecundity resulting from the orgy is amazing. A single sea slug can deposit 140 million eggs, strung together in spaghetti-like gelatinous ribbons, so the output from a ten-slug ring can be 1.4 billion eggs!

The only known vertebrate simultaneous hermaphrodites are found among the fishes. The black hamlet, a small sea bass living on coral reefs in the Caribbean, mates as a pair. One individual with ripe eggs initiates courtship. If the fish approached by the initiator is game to spawn, it cooperates. The initiator releases some eggs, and the partner fertilizes them externally. The partners reverse roles, and the other individual releases eggs to be fertilized. Then back to the first, who releases more eggs. This back-and-forth sex reversal usually happens several times before all the eggs of both individuals are fertilized. These fish engage in "egg trading": an individual gives up eggs to be fertilized in exchange for the opportunity to fertilize its partner's eggs. I use the word "opportunity" because sperm are much cheaper to produce than eggs. Thus, an individual should try to fertilize as many eggs as possible. What this all boils down to is that reproductive success as a male depends on its ability to reproduce as a female. Why? An individual can't fertilize someone else's eggs unless it gives up its own eggs to be fertilized.

Some simultaneous hermaphrodites—for example, most freshwater snails—can self-fertilize. This behavior is useful when snails become isolated from each other, such as when a snail sticks to the foot of a wading bird and is transported somewhere else where there are no snails of the same species. Some freshwater snails can self-fertilize for many generations without impaired vigor to their offspring.

THE SECOND STYLE of combining sexes is sequential hermaphroditism, or sex change over time. An individual that begins as a female and later becomes male is called a protogynous (from the Greek *proto*, meaning "first," and *gynos*, meaning "female") hermaphrodite. The reverse, an individual that begins as male and changes into female, is called protandrous (from the Greek *andro*, meaning "male").

Sex change is a normal part of some fishes' lives, occurring in at least fourteen of the approximately 435 families of fish. Eleven of these live on coral reefs. Sex change often depends on the social setting, and most often the change is from female to male.

Why should a female change sex? In some species, if a female switches sex at the right time, she will leave behind more offspring than if she stays female. Here's why: Generally a female's fertility is limited by the number of eggs she produces, and it doesn't matter whether she mates with one male or many. In contrast, a male is limited by the number of females he fertilizes, not by the number of sperm he produces. In many of the coral reef fishes, large males monopolize mating opportunities. For example, in many wrasses, damselfishes, angelfishes, and parrot fishes, older, larger males reproduce more successfully than do younger, smaller males because a large dominant male controls a harem. If this male disappears, within days or weeks the largest female becomes the dominant male, defending the territory and displaying sexual behavior toward her former harem mates. In this way, the female-turned-male will increase her fitness. And what about a fish that starts off male? All is not hopeless. A male born as a male must stay a male, but if he grows large enough, he might become a harem master.

Anemonefish, or clownfish, are unusual little orange, white, and black coral reef fish that eat, rest, and breed within the shelter of a large sea anemone's tentacles, which are lined with stinging cells that fire toxin into anything that touches them. Two species of clownfish are some of the un-usual fish where sex change goes from male to female. Clownfish have a completely different social system than the harem-operating coral reef fishes just described. Adult clownfish live as monogamous pairs, together with several subadults and juveniles. The young individuals are not offspring of the mating pair, since currents carry newly hatched fish far from their parents. For this reason, there is no inbreeding problem. This merry little band of adults, subadults, and juveniles rarely moves far from the safety of its anemone. In these species, the female is larger than and dominant over her mate, and both adults lord it over the smaller subadults and juveniles.

The adult male clownfish defends the anemone home against intruding adults, and he attacks any of the associated subadults that dare attempt to mate with his female. If the female disappears, the male transforms into a female. In less than four weeks, the former male-turned-female can lay fertile eggs. Who takes over the role of head male? You guessed it—the largest of the associated subadults. If both adults get eaten or otherwise disappear, the subadults grow quickly, the largest one becoming female and the next largest becoming male. There can be only one queen per anemone: she breeds and controls the sex of her anemone-mates. The subadults and juveniles are insurance that a replacement mate will be available if needed, without leaving the safety of the anemone to find one.

Slipper limpets (small marine snails) resemble tiny upside-down bedroom slippers. Although they are protandrous hermaphrodites, their mating behavior resembles that of the *Aplysia* sea slug, a simultaneous hermaphrodite. Slipper limpets spend most of their lives engaged in group sex, thus their scientific name of *Crepidula fornicata*.

These one-and-a-half-inch pale gray limpets live in shallow water from Nova Scotia to the Gulf of Mexico. After a brief period of floating about in the ocean, a larval limpet settles onto a rock or other substrate and prepares to metamorphose. Its drifting days are over. If the larva settles alone, it becomes a female. When another larva settles on the first limpet (limpets make great substrates and they're chemically attracted to each other), it becomes a male and develops a long tapering penis. Eventually another larva settles on the second. It becomes a male and the second individual gradually loses its penis and becomes a female. More individuals pile on. Adult slipper limpets usually live together in stacks of up to a dozen individuals. Each stack consists of a number of basal females, individuals in the middle undergoing sex change, and functional males at the farthest end from the point of attachment. How do males reach females at the other end of the stack? Recall that they have long tapered penises. An obvious difference between slipper limpets and *Aplysia* sea slugs is that the sea slugs form chains only for mating. The limpets live in their stacks and thus have continual access to their sexual partners.

An unusual reproductive system occurs in certain nematodes (roundworms). Although most nematodes have separate sexes, a few are hermaphroditic. In these species, sperm and egg production take place within the same gonad, called an ovitestis. Sperm are formed first, then the eggs follow. Technically, these roundworms are protandrous, but they don't cross-fertilize other individuals. Instead, the sperm are stored until eggs are produced. At this point, the sex cells get together. Voilà! Self-fertilization.

Sponges are simple sedentary creatures without symmetry and without organs. They are water-filtering systems that sit attached to a rock or plant and drink. Water enters the body through small pores, gets pumped through the body by a single layer of cells, and exits through a single large opening. Feeding cells, called choanocytes, filter out microscopic particles of food from the water current. Each choanocyte has a section of delicate tissue (the collar), which traps food particles, and a long threadlike structure (the flagellum), which whips around and creates water currents.

Sexual reproduction in sponges is a curious affair. Most sponges are sequential hermaphrodites. Some are protogynous; others are protandrous. Sex change may occur once or repeatedly throughout the sponge's lifetime. Cross-fertilization probably occurs in most species. Basically, one animal releases sperm into the water, and the current carries them to another sponge's body. A closer look, however, reveals that fertilization is anything but passive at the cellular level.

Sponges often release their sperm suddenly, in a smoky cloud. Some lucky sperm get washed onto another sponge. They don't just enter passively, however. The feeding cells (choanocytes) have a second function. They capture sperm and "ingest" them in the following way. A choanocyte encloses a sperm in a vesicle within its cell and then loses its collar and flagellum. It turns into a carrier cell and transports the sperm to an egg. After reaching its destination, the carrier cell fuses with the egg and effects fertilization.

With the unpredictability of seasonal sex changes amid an atmosphere of free love, reproduction in sponges could lead to provocative familial relationships. A sponge could fertilize its grandfather, father, or brother.

Likewise, it could receive sperm from its grandmother, mother, or sister during some future year.

A SPONGE PLANTED on a rock epitomizes an animal that benefits from being a simultaneous hermaphrodite, since it's unable to search for a mate. Earthworms, flatworms, and mollusks that don't move around much also benefit as simultaneous hermaphrodites since any partner will do. Certain coral reef fishes occur in low densities because coral reefs are often small and patchy in distribution. Thus, these fish might have a tough time finding mates of the opposite sex. Many deep-sea fishes also live in low densities. Thus, as you might expect, many of them are simultaneous hermaphrodites. Again, any partner will do. Simultaneous hermaphroditism is efficient: the sexual encounter of two individuals can result in both of them getting their eggs fertilized and in both being able to sow their seed.

What about sequential hermaphrodites? Why can't more animals change sex? Birds and mammals may have too rigid a system of sex determination to allow sex change. One reason sex change might not be more common in animals is that there may be a large physiological cost involved in switching the type of gametes an animal produces. There might also be behavioral disadvantages. For example, a female that switches to a male might be less successful in attracting mates than more experienced males in the population. All of this makes the existing sequential hermaphrodites that much more intriguing!

2 The Mamas and the Papas

TO CARE OR not to care? But that's not the only question. Who does the caring—mother, father, or both? Should care be fleeting or extensive? How does parental care benefit the young? What does it cost the parent? How flexible is parental care as circumstances change? Behavioral ecologists have pondered these questions for decades.

Most animals don't take care of their young. After they lay the eggs or give birth to the babies, the parents abandon their offspring, leaving them to face life's hazards alone. So why do some animals care for their young while most don't? In theory, the behavior of parental care evolves when the costs to the parent(s) are less than the benefits to the offspring of better survival. Costs are real. Caring for offspring requires time and energy, and it can be dangerous to parents. Also, in many cases, caring for existing young reduces further reproduction. Instead of incurring these costs, many animals simply reproduce frequently and/or produce many offspring at once; among the numerous eggs or babies produced, a few beat the odds to survive and reproduce in turn.

Other species, though, display extensive and sometimes amazing parental care. We ourselves care for our offspring longer than any other animal known, resulting in a strong parent-offspring bond that's not easily broken. We take for granted the care that birds and other mammals provide their young, but we're often amazed that some insects, spiders, fish, amphibians, and reptiles are devoted parents.

I'll describe some of the more unusual parental care stories from among the many thousands that naturalists have reported. The only birds and mammals I've included are the emperor penguin, platypus, and kangaroo. Instead, I'll emphasize invertebrates, fish, amphibians, and reptiles. First, we'll look at some animals that cater to their brood in harsh environments. Next, we'll consider three styles of care: tending progeny in stationary nest sites, carrying the young exposed on the parent's body, and carrying offspring in pouches. We'll end with a glance at parental sacrifice, including

some animals that provide the ultimate sacrifice: giving up their own lives
for their young.

SURVIVAL OF THE PAMPERED

I said to my husband one night, "I see our children as kites. You spend a lifetime
trying to get them off the ground. You run with them until you're breathless . . .
they crash . . . you add a longer tail . . . they hit the rooftop . . . you pluck them
out of the spouting . . . you patch and comfort, adjust and teach. You watch them
lifted by the wind and assure them that someday they'll fly. . . . Finally, they're
airborne, but they need more string and with each twist of the ball of twine, there
is a sadness that goes with the joy because the kite becomes more distant and
somehow you know it won't be long before this beautiful creature will snap the
lifeline binding you together and soar as it was meant of soar—free and alone."

    "That was beautiful," said my husband. "Are you finished?"

    "I think so. Why?"

    "Because one of your kites just crashed into the garage door with his car . . .
another is landing here with three surfboards with friends on them and the third
is hung up at college and needs more string to come home for the holidays."

ERMA BOMBECK, *if life is a bowl of cherries—what am i doing in the pits?*

Like most scientists, I like to think I'm on top of things in my laboratory.
Occasionally, though, I lose sight of what's happening right beneath my
nose—like the time I missed the amazing parental care behavior of some
poison dart frogs. (Their common name refers to the fact that some South
American peoples use the toxic skin secretions of several species to poison
darts for their blowguns.)

    In the late 1970s, a friend kindly brought me twenty strawberry poison
dart frogs from the rain forest of Costa Rica. Over the next two years, the
little red frogs gobbled up fruit flies, mated, and produced offspring—
poison dart frog style. The frogs courted and laid eggs
on damp leaves in the terraria. Once the tadpoles
hatched, they wriggled onto their mothers' backs. The
females carried their little black tadpoles to the water
caught between the leaves of the bromeliads I had
placed in the terraria. I assumed the tadpoles, like all
other poison dart tadpoles we knew about at the time, were eating algae
and other bits of organic matter. They were eating something, because the
tadpoles metamorphosed into tiny red frogs.

    I never paid much attention to the goings-on in the terraria, because I was too busy as an assistant professor writing lecture notes and advising students. All was going well, I figured. Studies of the tadpole-carrying behavior could wait.

Imagine my surprise when, in 1980, a German scientist published observations detailing parental care of strawberry poison dart frogs in captivity. The paper reported, for the first time, a frog feeding its tadpoles: after the female carries her tadpoles to bromeliads, she returns regularly and drops unfertilized eggs into the water for her tadpoles to eat. The tadpoles' entire diet consists of these eggs. Even more amazing, the mother and her tadpoles communicate with each other. The female peers into the water. If the tadpole is there, it swims close to the surface and repeatedly bumps its mother's hind end as she lowers herself into the water—all this as if to say, "Hey, Mom, I'm here and I'm hungry!" The female lays an egg. If there's no tadpole visible (perhaps it has died or metamorphosed), she hops away without leaving an egg. Clearly, I'd missed a great show in my terraria.

SO, WHY DO mother strawberry poison dart frogs carry their tadpoles to bromeliads and then provide meal service? Why don't they lay their eggs in ponds the way most frogs do? Think of it as an evolutionary experiment. Mortality is high for eggs and tadpoles developing in ponds. Often the breeding sites dry up before the eggs can hatch or the tadpoles can metamorphose. Ponds abound with voracious predators that stalk eggs and tadpoles. Over evolutionary time, some species of amphibians have come to breed on land. Parental care may improve the chances their offspring will survive in these new habitats.

Invertebrates and vertebrates alike have ventured into new habitats, but often they encounter challenges such as lack of food, insufficient oxygen, and water loss. Here we'll look at a sampling of animals whose parental care offsets those challenges.

As humans, we take breeding on land for granted; but for amphibians, land can be foreboding, as their eggs dry up quickly. Nonetheless, some frogs in tropical regions lay their eggs on land, in damp places such as beneath logs or under leaf litter. Some of these frogs give their eggs an extra moisture boost: one parent stays with the eggs and periodically provides a urine bath. Strange, indeed, but it works. One such example is Puerto Rico's national frog, a little golden-brown amphibian called the coquí, in which the male guards the eggs and moistens them by urinating on them. Hatchlings that develop in wetter nest sites are larger than those from drier nests. Thus, a father

coquí's urine is the magic potion to produce strapping offspring. In this species, the large eggs undergo a prolonged development of about three to four weeks and hatch as baby frogs—not as tadpoles.

An amphibian laying eggs out of water is one thing, but a fish? One small fish, the splash tetra from the Amazon Basin, is one of those odd evolutionary experiments that has worked, thanks to parental care. A shimmering gold-and-red male waits under overhanging leaves by the river's edge. When a female splash tetra is ready to spawn, she nudges the male with her head. The pair leaps out of the water, side by side. The female turns a somersault and lays an egg on the underside of an overhanging leaf, the male fertilizes it halfway through his somersault, and the pair dives back into the water. They continue to lay and fertilize eggs on the same and nearby leaves until the female has deposited all her eggs. She abandons both her eggs and her mate, leaving the male to keep the eggs moist until they hatch. He positions himself under the leaves, bends his body, and then with rapid tail flicks, splashes water up onto his offspring. The drier the weather, the more he flicks, taking a break only when it rains.

Even fish embryos in the water won't develop normally without enough oxygen. Parental care to the rescue. Some fish that live in still water aerate their eggs and fry (newly hatched young) by forcing water from their mouths or gill cavities over their offspring. Others, like the sticklebacks, fan their fins. Sticklebacks are small fishes, one to seven inches long, with distinctive spines on their backs. They live either in fresh or salt water, usually near coastlines, in the Northern Hemisphere. A male stickleback forms a tubelike nest from bits of plants and grains of sand, which he glues together with secretions from his kidney. He courts and entices a female into his nest, where she lays eggs. He then chases her away, fertilizes the eggs, and stays on to beat his fins over the eggs, providing oxygen-rich water until his young hatch days to weeks later.

Some salamanders also oxygenate their young. Two of these live in the rivers and streams of eastern North America: hellbenders and mudpuppies. Hellbenders, large salamanders growing to nearly thirty inches, are slimy and rather grotesque, with flat heads and grayish-brown wrinkled skin. A father hellbender rocks back and forth near his cluster of eggs on a river bottom, jostling them and increasing the flow of oxygen. Mudpuppies are six- to thirteen-inch aquatic salamanders with large external gills that resemble miniature ostrich plumes. A female mudpuppy attaches her cream-colored eggs one by one to the underside of a submerged log or rock. She stays with them until they hatch, vigorously beating her gills near the eggs. This action moves the eggs about and increases their exposure to oxygen. Another nurturing salamander dad is the Japanese giant salamander, the largest living amphibian, at about five feet. A male Japanese giant salamander continually agitates the string of eggs he's guarding for some forty to fifty days. This accomplishes the same end as the behavior of male hellbenders and female mudpuppies.

One way a frog can protect its young from hungry predators is to lay eggs in a water-filled tree hole. A male African tree toad calls from a hole high in a tree, advertising his prime nest site. A female climbs up the tree trunk in response to the calls. She lays twenty to thirty large yellowish eggs, resembling two strings of pearls, in the water. After laying her eggs, the female climbs back down and vanishes. The father hangs around for the next month. Life in a small tree hole is rough. Sometimes the water stagnates. The solution? From time to time, the father swims in place, kicking his feet rapidly in the stinky, stagnant water, aerating it like an air stone in a fish aquarium. Without his vigilance and calisthenics, the tadpoles would surely die.

Land is also an unusual environment for crabs, and only about 1 percent of all crab species live there. Females of one of those rarities, the bromeliad crab from Jamaica, care extensively for their young for some nine weeks. Less than one inch wide, these flat reddish-brown crabs live and breed in rainwater that accumulates in the leaf axils of large bromeliads growing on the ground of the mountain rain forest. The bromeliad water is low in calcium and dissolved oxygen, and it's acidic. Small though she is, the female crab actively removes leaf litter that accumulates in the nursery, increasing the surface area exposed to the air for gas exchange. She also pumps water through her gill chambers, maintaining a water current. Together, these activities increase the amount of dissolved oxygen. That's not all. Mothers dump empty snail shells into the water; as the shells dissolve, they add calcium carbonate and buffer the harshly acidic environment. The dissolved shells also provide calcium critical for her offspring's molting and development. Without the home improvements made by their industrious mothers, bromeliad crab larvae would die under the otherwise inhospitable conditions.

BROMELIAD CRABS ON Jamaica. Tree toads from Africa. Splash tetras from the Amazon Basin. Strawberry poison dart frogs from Costa Rica. All pamper their offspring under harsh environmental conditions. If fish can fan their fry, frogs can feed their tadpoles, and crabs can cater calcium for their larvae, what other behaviors have these animals "tried" unsuccessfully in the evolutionary past? What will be successful parental care behaviors 50 million years from now?

NESTS AREN'T JUST FOR THE BIRDS

There was an Old Man with a beard,
   Who said, "It is just as I feared!—
   Two Owls and a Hen,
Four Larks and a Wren,
Have all built their nests in my beard!"

EDWARD LEAR, from *A Book of Nonsense*

When I was a kid, this limerick conjured up vivid images of an aged Santa, his cottony white beard sprouting assorted bird nests and their inhabitants. Real birds' nests were even better, though, cradling pairs of brown-and-white speckled eggs or trios of turquoise blue eggs. I spent hours watching tireless parent birds carrying worms and grasshoppers back to their ravenous chicks. My youthful perception of nests as places where only baby birds lived, and where adult birds incubated, fed, and defended their young, was sadly naive. Many nonbirds construct nests as well, and nests are wonderfully diverse in architectural form, building materials, and function.

Consider function. Nests may create stable physical environments by controlling temperature or humidity, protecting against flooding, facilitating gas exchange and ventilation, or providing protection from predators. These cozy quarters may serve as places to store food, capture prey, or engage in courtship, and they may also function as nurseries. In some animals—for example, turtles and many insects—the parents abandon their nests once they lay their eggs. In others, such as most birds, the parents use the nest as a safe haven in which to warm, feed, or defend their eggs and developing young.

The following vignettes of nonbird nests and associated parental care reinforce the theme that diverse animals can accomplish the same ends—incubation, feeding, and defense—by similar means. Although in most cases a parent's presence serves more than one function, the following stories focus on one aspect of parental care. Again, I've singled out some of the more unusual behaviors associated with nests and eggs.

SOME PARENTS USE a nest as a place to warm their eggs, since eggs can't regulate their own temperatures. Birds incubate their eggs with their own body heat. Most snakes that brood eggs simply coil around them and guard them. They have no way of keeping the eggs warm. Some pythons are exceptions, however.

When ready to build a nest, a female Australasian diamond python burrows into sandy soil and makes a shallow depression. She pushes leaf litter into the basin and forms a mound into which she lays her eggs. Over the following eight to ten weeks, she coils around her eggs and keeps them warm by contracting her muscles, which produces metabolic heat. Amazingly, she keeps the eggs an average of $16°$F warmer than the surrounding air during the day and up to $32°$F warmer at night. This shivering activity is costly, however. During the time she incubates her eggs, a female may lose as much as 15 percent of her body mass.

The heavy-bodied Asian rock python uses the same warming trick. These snakes can reach over fourteen feet. They feed on deer and even adult leopards. Unlike most snakes, though, they're caring mothers. Like the diamond python, a female Asian rock python coils around her eggs and contracts her powerful muscles to generate heat, maintaining the eggs at about $86°$F until they hatch eight weeks later. She tightly closes her body coils when the air is cool and loosens her coils when the air is warm. The cooler the air temperature, the more she twitches.

MANY ANIMAL PARENTS provide food to nesting young that can't scare up their own meal. More than 70 percent of all birds (over five thousand species) lay eggs that hatch into naked, blind chicks—demanding young that must be fed constantly. But we all know about birds, so let's look at some other animals that lay their eggs in nests and then feed, or at least provide food for, their helpless babies.

Some animals make nests right in their food source. Burying beetles, about one to one and a half inches long, have a keen sense of smell that leads them to dead animals. A pair of burying beetles enters a carcass of a small bird or small mammal and checks it out to see if it will make a suitable home for their young. If the beetles approve, they chew the carcass and mold it into a roughly spherical mass. After excavating a burrow underneath, they push their prize belowground and then seal up the burrow from below. Now isolated with the ball of decaying flesh, the female eats out a crater-shaped depression, into which she lays her thirty or more eggs. After the young hatch, they sit in this nestlike crater and beg, just like baby birds. They stretch upward, mouths ready to engulf the liquefied carrion their parents regurgitate.

Other animals provision their nests with food so that when their eggs hatch, the young will have food even though the parent is long gone. Cicada killers, showy black-and-yellow two-inch wasps, are among the largest wasps in North America. Females dig elaborate underground burrows, each with about sixteen brood chambers. Once finished digging, the females lay in the food supplies: cicadas they have stung and paralyzed. Cicadas can weigh three times the mass of a wasp, so the cicada killer sometimes drags her catch along the ground, up the trunk of a tree, and then launches herself into the air and flies with her victim down toward her burrow. Once the cicada is safely ensconced, the wasp lays an egg on it. A few days later, the egg hatches and the larva devours its food.

There's more than one way to stock a nest. A female pipe-organ mud-daubing wasp builds a mud nest on a sheltered surface such as under a bridge or eave of a house. Each tubular nest contains two to five brood cells, each of which will house one offspring. Once her nest is complete, the female flies out in search of a spider, which she stings and paralyzes. She squeezes the spider's body, causing it to regurgitate its stomach contents. Placing her mouth to the spider's mouth, she steals a meal for herself before carrying the spider back to her nest. There she delivers the paralyzed prey to the first cell; she stashes additional spiders in the cell over the next day or two. She then lays a single egg on top of the food cache and seals off the cell with mud. No rest for the weary. Immediately she begins to pack the next brood cell with spiders, repeating the process until she has laid an egg in the last cell. Each egg hatches in about two days, and the hungry larvae devour all their food in about one week. With no more food available, the larvae spin cocoons around themselves and stay inactive until they pupate and later, as adult wasps, chew their way out of the nest.

Including mammals in a discussion of laying eggs and feeding young from nests might seem odd. But there is one: the duck-billed platypus from Australia. A female platypus digs an elaborate thirty- to sixty-foot burrow in a stream bank. At the end of her tunnel, she excavates a nest chamber. She then spends the next two weeks gathering leaves and grass stems, which she carries back to her burrow in her leathery ducklike bill.

Once the nest has met with her approval, she plugs the tunnel in several places between the entrance and the nest chamber. Now she's safe from predators and will spend the next few weeks in darkness and self-imposed seclusion. She lays two eggs, each less than an inch long, and then takes an extended nap while she incubates her eggs by holding them pressed to her belly with her beaverlike tail. After ten to twelve days, thumbnail-sized underdeveloped baby platypuses hatch by ripping open the leathery shells with their "egg teeth." The mother snoozes while her babies develop. Throughout these three or more weeks spent in darkness, the female has been metabolizing fat stored in her tail, but eventually hunger overtakes her and she leaves the burrow in search of worms, beetles, and crayfish. Once the babies' mouths develop, they, too, become hungry and begin to nurse. Platypuses have mammary glands hidden under the skin of their abdomens, but they don't have nipples. Instead, milk seeps out from the glands onto the mother's fur, and the babies lap it up with their tongues.

I LEARNED EARLY on that birds keep close tabs on their nests, defending their young from harm. But an angry robin can't hold a candle to a male gladiator frog. These Central and South American frogs have large sharp spines on their thumbs that they use to gouge their opponents' eyes and ears—hence, the common name. A male gladiator frog's defense is crucial for his eggs' survival. Pivoting on his belly in the soft mud, a three-inch male pushes debris out to the side with his palms and flings mud backward with his feet to excavate a basin eight to eleven inches in diameter. The basin fills with rainwater and serves as a nest site from which the male will spend countless hours calling to attract a mate. His call, *tonk-tonk-tonk*, sounds like someone monotonously beating on a chunk of wood. It doesn't work wonders on human ears, but if the male is lucky, a female gladiator frog is enamored and lured to his nest. After laying two to three thousand eggs on the water surface, the female takes off, but the

father's work has only just begun. His job is to protect their eggs. If the surface tension is broken, the eggs will fall to the bottom and die from lack of oxygen. What would break the surface film? Other males. Why would they do that? If an intruder male kills other individuals' offspring, he will increase his own reproductive success relative to other  individuals in the population. So, the new father patrols his nest for the first couple of days. If another male comes too close, the father attacks. The two wrestle in a bear-hug hold, growling and hissing, sometimes with dire consequences: gouged bodies.

In many species of spiders, the mother dies soon after enclosing her eggs in a silken egg sac—a spider's style of nest. Some spiders, though, live on and defend their offspring. Consider the three-quarter-inch green lynx spider. Soon after mating, the male dies. The female continues to eat for a few weeks and then lays eggs that she enshrouds in a silken sac. Using silk, she secures her egg sac and its valuable contents to a leaf. Throughout the next six to eight weeks, she guards her sac from predators such as ants, on occasion even spitting venom at intruders. Eventually the spiderlings hatch. Once they emerge, the spiderlings hang around the sac for several days. Still the mother guards and defends her young. Only after the young have dispersed is her work complete. She is now free to die, which she soon does.

In 1990 Giselle Mora, then a graduate student at the University of Florida, discovered that a harvestman from Panama (*Zygopachylus albomarginis*) provides paternal care—the first reported case of parental care by a male arachnid. (Harvestmen, commonly called "daddy longlegs," belong to the class Arachnida, along with spiders, mites, scorpions, and their relatives.)

Males of this particular harvestman species construct cuplike nests on standing trees, fallen logs, or rocks. They collect their nest materials—mud and bits of tree bark—by scratching tree bark with their chelicerae (front appendages). They roll the material into pinhead-sized irregular balls and then carry the balls back to their chosen nest site. A male usually begins nest construction by adding salivary secretions to the mud-and-bark ball. He then presses the ball onto the substrate and makes a circular floor about one and a quarter inches in diameter. Over the next day or two, he collects and carries additional balls and applies them to the edges of the floor to make walls. He works around in a circle, adding new balls to old, gradually building up the height of the walls to a little more than a quarter of an inch.

Giselle found that females approach nests, initiate courtship with the nest owners, lay eggs in the nests, and then leave the males to care for the eggs. At a given time, a female lays from one to four eggs in a nest. She may stay near the nest and mate again with the male at a later date, or she may find a different male to seduce. Males usually guard eggs from more than one female, and the eggs are at different stages of development. A male protects an average of about twelve eggs at a time.

A father harvestman's life is a busy one. He removes fungus from the nest by eating it. He also repairs the nest when it becomes damaged by rain, wind, or other animals, or when it dries out and the walls collapse. In addition, he guards his eggs and newly hatched juveniles against predators. He picks up individual ants and tosses them away from the nest, though when an ant swarm invades, he throws in the towel and abandons his young. Other harvestmen, male and female, are the most common predators; a father will chase and bite any individual that gets too close to his nest.

Fish, too, can vigilantly defend their young. Have you ever noticed the gaudily colored blue-and-red fish with showy fins swimming around in those little glass goblets in the pet store? These Siamese fighting fish (also called Betta, after their scientific name *Betta splendens*) are kept in solitary confinement because of their well-known pugnacity. In spite of their champion fighting skills, males have a gentler side: they're good fathers. Like female green lynx spiders, they guard their offspring until the young disperse. The nest is different, though—not silk but rather bubbles.

A male Siamese fighting fish gulps air and then blows mucus-covered bubbles toward the water surface to form a frothy nest. The mucus strengthens the walls of the bubbles, lengthening their life span. A willing female swims under the bubbles, and the male fertilizes her four to five hundred eggs as she releases them. Both parents then swim after the downward-drifting eggs, slurp them into their mouths, and spit them up into the bubble nest. Once all the eggs have been fertilized and spat into the nest, the male drives his mate away and ousts potential predators. If any eggs fall out of the nest, he catches them in his mouth and spits them back up to join the others. The eggs hatch within two days, but the young hang from the bubbles awhile longer, absorbing the remaining yolk, all the while defended by their father. Only when the young swim free from the nest does the male end his vigilance.

A female American alligator guards her eggs in the nest and afterward cares for her babies until they can defend themselves. The female constructs a nest about three feet high and up to seven feet across by heaping vegetation and soil into a mound. She lays twenty to sixty eggs in the center of the nest, where the vegetation is moist. Heat from the sun and the rotting vegetation incubates the eggs over the next nine weeks or so, and the mother hovers close by. She threatens all intruders, including humans, that come between her and her nest. As they begin to break out from their eggshells, the babies peep, attracting the mother's attention. She rips open the nest and frees them. Carefully, she nudges one or two babies into her toothy mouth and carries them to a nearby pond, where she releases them. Imagine riding to safety between a set of teeth like that! Back she goes, to transport more young to water, continuing until all her babies are safely ensconced in their new home. Throughout the next year or two, she stays with her babies, who need only peep and she comes running (or swimming).

As a rule, nonreproductive worker ants protect baby ants in nests. One of the most bizarre twists, though, is seen in the weaver ant, a species in which babies help make their own nest. Weaver ants from Africa, Asia, and Australia construct tentlike nests in the treetops with large rooms defined by walls, floors, and ceilings for their queen and her brood, their food supplies, and themselves. The workers begin by grasping each other in living bridges to pull two large leaves together. Then, another team of worker ants uses larvae that are almost ready to pupate as living shuttles to weave the leaves together with strands of silk. The workers gently hold the larvae in their mandibles and move their "tools" back and forth between two leaf edges as the larvae release silk threads from glands located below their mouths. The larvae give up all their silk in the process. So what do they do, no longer able to spin cocoons around their bodies to protect themselves while they pupate? They don't need cocoons; they're safe in the shelter of the nest they've helped to make! And whenever alien ants or other potential child molesters approach the nest, workers boil out, ready to defend the young by sinking their sharp, powerful mandibles into the intruders.

NESTS. CREATIVE INVENTIONS not just for the birds! Nests range from simple to elaborate, from small to immense. Some are networks of threads, rolls of leaves, or woven twigs. Others are extensive burrows with branching corridors and multiple exits, or tunnels with brood chambers. Building materials may come from the animals themselves or be picked up from the environment. Spiders spin silk. Some termites, beetles, and other insects use their own or others' feces. Some fish use mud or vegetation, or air and mucus. A few amphibians form mud into nests; others whip a combination of air and mucus into meringue nests. Some reptiles dig nests in the mud or sand or pile vegetation into a heap. Nest-building mammals use twigs, grass, leaves, or mud. From the safety of a nest, a parent can lash out at predators, regurgitate or bring unadulterated food for its babies, and keep its young warm.

BABIES ON BOARD

The whole [beaver] family can be seen rolling, porpoising, and somersaulting. It is possible to spot a father beaver swimming with great speed across the lake

with one of his kits hanging on, then turning around and swimming back, sometimes out of sheer exuberance.

JEFFREY M. MASSON, *The Emperor's Embrace*

We humans also sometimes carry our toddlers piggyback, just like mother
and father beavers carry their kits. Many other animals carry their young
exposed on their bodies for a much longer time. There are many advantages to portable young. The parents can go about their business, including eating, without leaving the babies home alone. If a predator approaches,
the parent can attempt escape with the young instead of fighting and risk-
ing death, not to mention leaving behind orphaned babies. If it gets too
hot or cold, wet or dry, the parent can move its young to a better spot.

DURING A SABBATICAL year from the University of Florida, my
family and I lived in the mountains of Costa Rica, in the community of
Monteverde. Our daughter, Karen, was a mere six weeks old when we began our adventure, and Monteverde was a wonderful place for naive first-
time parents to raise a child—save for the house scorpions. Shortly after
arriving, I spied a large black, sinister-looking scorpion scuttling across
the floor near Karen's crib. The scorpion carried a load of tiny whitish babies on her back. Freaking out at the thought of her crawling into Karen's
crib, I instinctively stomped on the creature. My next reaction? Guilt. I
was appalled at myself as a biologist. Like me, this scorpion mother was
just protecting her brood. I vowed then that I would try to live in peace
with these intruders—or at least to relocate them somewhere far away
from my own offspring.

Scorpions give birth to live young. In many species, the female folds
one or more of her pairs of legs under her genital opening while giving
birth. That way, the babies fall gently into this "birth basket" instead of
plummeting onto the hard ground. Soon after
birth, the young clamber up their mother's legs
onto her back. Depending on the species, the
young may spread out in a single layer and ori-
ent themselves in an organized manner, or they
may pile up in haphazard fashion several individuals deep. Either way, the babies ride piggyback
until after their first molt, which usually happens a few days to two weeks
after birth. The young don't eat while on board. After they molt, the young
leave the mother and scatter in search of their first meal, a small insect or
spider.

A female wolf spider encloses her eggs in a silken sac, which she drags behind her attached to her spinnerets (the organs that spin silk). She fiercely defends her eggs against predators, and she sunbathes to warm her eggs. Occasionally the mother detaches the sac and examines it, presumably to monitor her eggs, and then reattaches it. About a half day before they emerge, baby wolf spiders become hyperactive. This activity stimulates the mother to detach the sac once more, but this time she slits it open with her fangs, an essential step that the babies can't do themselves. After making a slit, the mother reattaches the sac to her body. Eventually the babies stream out of the sac and pile up onto their mother's back, where they'll ride for the next week or more.

Just as with scorpions and wolf spiders, insects that carry their young presumably enhance their offspring's survivorship during the early vulnerable stages. My favorite egg-carrying insect dad is the giant water bug—my favorite in spite of the fact that I once watched a huge Argentine giant water bug eat one of my study animals, a narrow-mouthed toad. Giant water bugs are scary-looking: up to four inches in some species, with strong beaks, grasping front legs, and voracious appetites.

Male giant water bugs carry their eggs on their backs and protect and aerate them for fifteen to thirty days until the nymphs hatch. Using bodily secretions, the female glues the eggs to her mate's back. He doesn't allow her to do it all at once, however. Male giant water bugs are control freaks.

A male repeatedly interrupts his mate while she's laying eggs and forces her to copulate. After each copulation, the female mounts her mate and lays one or more eggs until he forces her to copulate again. In this way, the male is assured that he is indeed the father of the eggs he will have to carry. In all, he may demand more than a hundred copulations over the course of two days. A male can carry as many as 140 eggs, all neatly packed together.

Giant water bugs usually prefer deeper areas of a pond. When males are carrying eggs, however, they cling to plants near the water surface, where the eggs are exposed to more oxygen. Egg-laden males of some species vigorously stroke the eggs with their hind legs, circulating the water to keep the eggs aerated. Males of other species methodically rock back and forth. Eventually the babies hatch and swim away.

Most crustaceans (crabs, lobsters, crayfish, isopods, and their relatives) carry their eggs on their abdomens, where they protect the eggs until they hatch, weeks to months later. Some female lobsters lay up to a hundred thousand eggs at a time and carry them for up to a year. After their young hatch, these weary mothers take a year off before breeding again.

Frogs exhibit diverse reproductive behaviors, with about 10 percent of the four thousand or so species providing some type of parental care. Some display unique ways of carrying their eggs or tadpoles.

Some frogs mimic the giant water bug technique, though instead of pawning the care off onto the male, the female herself carries her eggs exposed on her back. Mucus glands from the female's back secrete a glue that fastens the eggs onto her skin. One of these is the egg-brooding horned treefrog from Ecuador: a light golden-brown frog with a somewhat flattened body, huge triangular head, wicked-looking bony "horns," spines of skin poking up from the eyelids, and a fleshy proboscis extending from the snout. A female carries about thirty eggs on her back—eggs that skip the tadpole stage and instead undergo development within the capsule until they hatch directly as little froglets. We don't know much about the natural history of this South American frog, but presumably it's similar to a closely related species from Panama. In the Panamanian species, when the froglets hatch, each is attached to the mother's back by a pair of "gill stalks." These stalks extend from the froglet's throat region to the egg capsule still stuck to the mother's back. The mother carries her young around, each connected by this pair of miniature leashes, for an unknown period of time. Eventually the stalks slough off and the froglets tumble to the ground.

Midwife toads from Europe lay their eggs on land, but the eggs hatch into tadpoles that must develop in water. The father accomplishes this transportation feat in a remarkable way. After the pair lays and fertilizes strings of twenty to sixty eggs, the father thrusts his legs through the egg mass. The sticky egg strings adhere to him, and he stumbles around for the next few weeks with the eggs entwined around his thighs and waist. Periodically he dips into shallow water, ensuring the eggs don't shrivel up and die. When the tadpoles are nearly ready to hatch, they flip about in their egg capsules. The wriggling of twenty to sixty embryos likely tickles the male's body and stimulates him to hop to a pond. Some of the tadpoles burst forth from their capsules and swim off. The father hops to another pond where more tadpoles hatch. Perhaps by scattering his tadpoles, he "hedges his bets" (the "don't put all your eggs in one basket" principle). Even if some ponds dry up, offspring in others may survive. The male continues to scatter tadpoles until finally his legs are kid-free.

Poison dart frogs and rocket frogs from Central and South America lay and fertilize about four to thirty eggs on land, inside a curled leaf or some other protected moist site. Once they hatch, the tadpoles must get to water. Depending on the species, either the mother or the father provides piggyback transport. In some species, the parent carries the tadpoles one by one to water-filled tree holes or the water that accumulates between the leaves of bromeliads. In other species, all the tadpoles ride at once, usually to a shallow stream. In some cases, the tadpoles ride for only a few hours before being dumped; in others, for several days to over a week. And then there's always one that has to be different. Tadpoles of Degranville's rocket frog hatch with huge bellyfuls of yolk—enough to complete development while riding on their father's back. Instead of transporting his tadpoles to water, the father carries his young around until they hop off on all four feet.

NOW YOU KNOW what scorpions, wolf spiders, giant water bugs, crustaceans, and some frogs (as well as many other animals not mentioned here) have in common: by carrying their babies, they all gain freedom to go about their daily business without leaving the kids behind to get into trouble. Nests work fine for some animals, but for others the benefit gained by carrying babies on their bodies offsets the cost in time and effort and the possible risk to life and limb. In some cases, such as the poison dart frogs and rocket frogs, the piggyback transport is critical to move the newly hatched offspring from land to water.

A POUCH FULL OF MIRACLES

We find a Strange Animal among us. An animal of whom we have never even heard before! An animal who carries her family about with her in her pocket!

A. A. MILNE, *The World of Pooh*

The scene, of course, is the Hundred Acre Woods, and the "Strange Animal" is Kanga, carrying around little Roo in her pouch.

Kangaroos remind me of an experience I had in Dubrovnik, Yugoslavia (now Croatia). My six-month-old son was sleeping in a brown corduroy baby carrier strapped to my chest as I strolled along a busy street. A young Gypsy woman with a child sleeping in a multicolored sling draped across her chest walked toward me. With flamboyant gestures, she pointed to her baby, then to mine, and back to hers, grinning at the idea we were both carrying our babies in pouches.

IF YOU'RE GOING to carry your babies, you might just pile them on. But a pouch might be one step more secure, preventing them from falling off. Pouches occur in a variety of animals and at various locations on their bodies, from the mouth to the underside of the body. Let's look at a sampling of these animals, starting with those sporting abdominal pouches.

One to two days before giving birth, a red kangaroo mother holds her pouch open with her forepaws, sticks her head inside, and licks it thoroughly. When she senses her time has come, she leans against a tree, extends her hind legs, rotates her pelvis forward, positions her tail between her legs, and gives birth. Although the baby's front legs are fully formed at birth, its hind legs are mere stumps. Less than a minute after being born, the lima-bean-sized baby begins its arduous journey to the pouch. Using its forelimbs to swim through its mother's fur, guided by its sense of smell, the baby clambers into the pouch within three minutes. Mission accomplished, the newborn quickly latches onto one of the four nipples, which soon swells to fill the baby's mouth. The baby, or joey, will stay continually attached to the nipple for the next 120 to 130 days. This strong grip keeps the joey from bouncing from one side of the pouch to the other as its mother bounds across the landscape.

At about 150 days, the joey pokes its head out of the pouch for a first peek at the outside world. In another forty days or so, it leaves the pouch for the first time—usually by falling out. Later on it perfects the technique of diving out headfirst. As the days pass, the joey spends more and more time outside the pouch. Whenever the mother senses danger, however, she bends forward and holds the pouch open with her forelimbs. In an amazing acrobatic display, the youngster kicks off with its hind legs and does a complete somersault so that it lands in the pouch face forward. Imagine a human toddler mastering this feat to reenter a baby carrier or a Gypsy's sling!

After another fifty days or so, the youngster leaves the pouch permanently, but for another few months periodically sticks its head inside to suckle on its personal teat. It will not be weaned until it's nearly a year old. Toward the end of this time, the mother increasingly resists the youngster's pestering and finally refuses the nipple.

That's the story for one joey, but he or she is rarely an only child. Within one to three days after giving birth, a female kangaroo often mates again and another fertilized egg settles into her uterus. Female red kangaroos seem to be nothing but baby-making machines, but there's more to it than that. Red kangaroos live in the dry shrub and grasslands of central Australia. Some years severe droughts kill the young kangaroos, so it's advantageous for a mother to give birth again quickly if her baby dies.

Flexibility is the key. The fate of the egg that's fertilized when the mother mates again just after a baby's birth depends on what happens to that baby. If the recently born baby has reached the pouch and latched onto a teat, the newly fertilized egg develops only to the blastocyst stage (about seventy to one hundred cells) and then stays in developmental dormancy until it receives the signal to continue development. The blastocyst can remain dormant in the uterus for several months. If the newborn baby hasn't made it to the pouch or if it dies while in the pouch, the female's body releases a pulse of the hormone progesterone. This signals the blastocyst to continue to develop, and about thirty-three days later the "replacement baby" will be born. The same happens once a mature joey is about to leave the pouch permanently and is nursing less: the mother's hormones kick in and the fertilized egg begins to develop. Thirty-three days later she'll give birth to a new baby, which will scramble up its mother's fur and attach to a nipple in the pouch.

A female red kangaroo's life is demanding. At any one time, she may provide a safe haven in her uterus for a dormant embryo, milk for a joey attached to one of her nipples, and milk for her oldest youngster, who pokes its head in the pouch from time to time demanding its fair share. In addition, this "teenage" joey requires considerable attention as its mother teaches it to graze for food and avoid predators such as dingoes and eagles. Most amazing is that red kangaroo mothers produce two types of milk for their nursing babies. The one attached to the teat in the pouch receives low-fat milk, whereas the active teenager gets higher-fat milk.

Kangaroo pouches are unique, but the concept of carrying the family in a front pocket is not. Male seahorses immediately come to mind. After an extended courtship, a female uses her ovipositor to lay from ten to many hundreds of eggs in her mate's pouch. As the eggs enter the tiny pouch opening, the male fertilizes them. He carries around his load, essentially in a state of pregnancy. Each embryo receives oxygen from capillaries in the male's tissue. The embryo also receives nourishment: the hormone prolactin from the male causes the egg membrane to break down, yielding a nutritional fluid. Eventually the eggs hatch, and the baby seahorses stay put until they're strong enough to swim actively. At that point, the male expels his young by pumping and thrusting, a labor that may last several days. The young are on their own to fight the currents and flee from predators.

Some starfish broadcast their eggs into the open ocean, but others brood their young in pouches on the ventral sides of their bodies. One brooder is the intertidal six-armed starfish *Leptasterias hexactis,* studied extensively near Friday Harbor in Washington State. In this species, the female forms a brood chamber by arching her central disk away from the substrate (usually the underside or side of a rock) and then pulling the tips of her arms toward the center of her body. She deposits her eggs in this pouch and broods them there for about two months, until the young metamorphose into small starfish. Unless disturbed, a female stays attached to the rock throughout this time. While brooding, a female does not eat; she metabolizes stored energy. Rather than sitting passively, she continually manipulates the eggs to clean them of debris and to aerate them. Brooding also protects the developing embryos from exposure to extreme environmental conditions and helps keep them safe from predators, such as other starfish. Experiments have shown that brooding in this species is critical. Without brooding, none of the experimentally isolated embryos survived for more than five days. The embryos soon became covered with debris and infected with bacteria.

Remember from chapter 1 Steve Shuster's harem-forming marine sow bugs (*Paracerceis sculpta*) that mate inside breadcrumb sponges? Not only is their mating system unusual; their parental care system is also a bit peculiar.

Several days after one of these female sow bugs enters a sponge, she begins her sexual molt. First she loses the rear half of her cuticle, which exposes her genital pores and leaves her sexually receptive. She mates repeatedly over the next twenty-four hours, then completes her molt by shedding the front half of her cuticle and losing her functional mouthparts. She lays her eggs in an internal pouch on the underside of her body. The eggs develop in this pouch over the next two to nine weeks. The warmer the temperature, the faster the embryos develop.

Brooding is hard on the mother sow bug. She doesn't eat during this time, both because she doesn't have working mouthparts and because when her pouch is full, it occupies almost all the space within her body cavity. Steve suggests that because a brooding female loses so much muscle mass and depletes so much of her stored fat, it's likely that the embryos gain nutrients from the body fluids of their mother. Eventually the juveniles leave the pouch as small mobile individuals. The female dies within the next two weeks. She breeds only this one time.

SOME FROGS CARRY their offspring in pouches on their backs. Forty or so species of marsupial frogs, named for the egg-carrying pouch on the female's back, live in the New World tropics: the Andes plus Panama and eastern Brazil. Pouches come in many forms, from relatively open ones with an elongate slit running down the middle of the female's back to closed pouches with either wide or narrow openings just forward of the female's cloaca (the common opening through which intestinal and urinary wastes and eggs pass).

In the high Andean *paramó* (tropical alpine habitat) of Ecuador lives the mottled green-and-brown Riobamba marsupial frog. The female has a closed pouch with a small circular opening on her lower back. In order to get the 130 or so eggs into the incubator on her two-inch-long body, the female executes some acrobatics and the male guides the eggs while clasping his mate. By twisting her back legs, she raises her cloacal opening above the pouch opening. With his feet, the male smears fluid containing sperm onto his mate's rear end. As the female releases eggs, the male pushes them with his feet through the fluid and into the pouch opening. The eggs eventually hatch into squirming tadpoles inside the pouch, and the female waddles to a pond or puddle. Once in the water, she pokes a toe into her pouch and stretches the opening. With her hind feet she scoops out the tadpoles, digging deep to coax out the recalcitrant ones stuck in the far reaches of their incubator. The freed tadpoles develop in the water until they metamorphose into froglets.

Another impressive egg-carrying frog is the one-inch reddish-brown Puerto Cabello marsupial treefrog from the cloud forests of Venezuela. The female has a thin pouch on her back with a slitlike opening running down its length. Prior to mating, the male sticks his feet into the female's pouch and kicks, stretching the pouch in preparation for receiving the nine or so large eggs. As the female pushes out the eggs at intervals of less than one minute, each lands on her back, where the male grasps it between his heels. He rotates each egg several times and touches it to his cloaca, fertilizing the egg about four seconds after the female has extruded it. Still grasping the egg between his heels, he pushes it into the pouch. After fertilizing the last egg, the male releases his mate and wanders away. The female goes about her business for the next twenty-five days or so, carrying the eggs in her pouch. She's so small that the large eggs nearly double her weight. When the eggs are ready to hatch, the female lowers her body into the water accumulated in a bromeliad, and the two halves of the pouch separate. Tadpoles endowed with well-developed limbs and lots of yolk plop into the water. The well-developed limbs give these tadpoles a head start in their aquatic nurseries, and the yolk provides food in their otherwise food-limited environment.

ONE AUSTRALIAN FROG carries its young in pouches located along the sides of its body. Hip-pocket frogs, less than half the length of your little finger, lay and fertilize about eight eggs on land. The female guards her eggs, hidden in the leaf litter, while the male remains close by. Less than two weeks later, the father takes over babysitting duties. A male doesn't actually have pockets in his hips, but he has the closest thing to it: a pouch along each side of his body, extending from just behind his armpits to his groin. The pouches are incubation chambers large enough to house his tadpoles. On about day 11 or 12, he climbs into the small egg mass. His gyrations rupture the jelly capsules, and the tadpoles pop out. They wriggle about until they locate the pouch openings at the groin end, and in they squirm. The tadpoles' abundant yolk supplies will nourish them until they transform into miniature hip-pocket frogs and hop back out.

AND THEN THERE are Darwin's frogs. A few years ago I had the good fortune to receive funding from the National Geographic Society to study parental care in Darwin's frogs. (Charles Darwin discovered these frogs while exploring the forest around Valdivia, Chile, in 1835, thus the common name.) These one-inch brown or green frogs with Pinocchio-like snouts live only in Chile and Argentina.

A male Darwin's frog peeps softly and attracts a female, who follows him to his hiding place—a protected moist spot under leaf litter or in a small cavity. After courtship, the female lays her eggs, usually fewer than ten, and the male fertilizes them. The father stays nearby for the next three weeks or so, and when the embryos begin to rotate in their translucent egg capsules, he gulps them into his mouth. There they slide into his vocal sac (the pouch that fills with air and functions as a resonating chamber when the male calls). Hormonal changes during the breeding season cause the vocal sac to enlarge, and it becomes an incubation chamber. Once inside the pouch, the eggs soon hatch. And there the tadpoles grow and develop, absorbing moisture and nutrients from the vocal sac lining over the next fifty to eighty days. I expected that brooding males would be hard to find, that they wouldn't expose their precious cargo. Not so. Brooding males sat fully exposed, soaking up the sunshine and snacking on insects. After metamorphosing into froglets, the young crawl out into their father's mouth. He gags, and his babies hop out. The froglets are dark gray when they emerge, but eventually their skin takes on the lovely greens and browns typical of their species.

THE MOUTH ITSELF makes a good nursery for some fish. Representatives from ten different families of fishes use their mouths as incubation pouches: the parent holds the eggs and offspring in the mouth or in the gill cavities, where they develop. During this time, the parent continually makes churning movements in its mouth, ensuring the young have a constant supply of oxygen and are kept free of waste material. In some species, the fry eat tiny food particles flushed into their incubators. Often the offspring stay in their parent's mouth until they're fairly well developed. Examples of mouth-brooders include sea catfishes, cichlids, cardinal fishes, and bonytongues. In some species, after the parent releases the young, they stay close to the parent, fleeing back to the mouth when danger threatens—behavior reminiscent of a joey somersaulting into its mother's front pocket.

In many mouth-brooders, the female provides the care—but not always. In one cichlid species from Lake Tanganyika in Africa, the mother broods the fertilized eggs in her mouth until after they hatch. Once the young outgrow her mouth, she transfers them over to the father, whose mouth is larger. There's a catch, however. She must identify the father from all the other fish in the lake. If she points a fin at the wrong guy, he just might eat the young instead of brood them.

Konrad Lorenz, the famous Austrian naturalist and animal behaviorist, describes an amazing observation of a captive male jewel fish (a striking red-and-blue cichlid) in his book *King Solomon's Ring:*

I came, late one evening, into the laboratory. It was already dusk and I wished hurriedly to feed a few fishes which had not received anything to eat that day; amongst them was a pair of jewel fishes who were tending their young. As I approached the container, I saw that most of the young were already in the nesting hollow over which the mother was hovering. She refused to come for the food when I threw pieces of earthworm into the tank. The father, however, who, in great excitement, was dashing backwards and forwards searching for truants, allowed himself to be diverted from his duty by a nice hind-end of earthworm. . . . He swam up and seized the worm, but, owing to its size, was unable to swallow it. As he was in the act of chewing this mouthful, he saw a baby fish swimming by itself across the tank; he started as though stung, raced after the baby and took it into his already filled mouth. It was a thrilling moment. The fish had in its mouth two different things of which one must go into the stomach and the other into the nest. . . . The fish stood still with full cheeks, but did not chew. . . . For many seconds the father jewel fish stood riveted and one could almost see how his feelings were working. Then he solved the conflict in a way for which one was bound to feel admiration: he spat out the whole contents of his mouth. . . . Then the father turned resolutely to the worm and ate it up, without haste but all the time with one eye on the child which "obediently" lay on the bottom beneath him. When he had finished, he inhaled the baby and carried it home to its mother. (pp. 42–43)

PERHAPS MOST BIZARRE of all, in 1974 Australian biologists reported a new parental care behavior: brooding young in the stomach. First
in the animal kingdom, this claim to fame belongs to a little stream-dwelling
frog called the southern gastric brooding frog, known only from southeast
Queensland. The female swallows her twenty-one to twenty-six fertilized
eggs or newly hatched tadpoles, and there they develop, in her stomach, until six to eight weeks later when her esophagus dilates and she belches up
little froglets. While developing in the mother's stomach, the tadpoles feed
exclusively on their large yolk stores. The mother doesn't eat during this
time, but why doesn't she digest her babies? The young secrete a substance
called prostaglandin E_2, which inhibits acid secretion in their mother's
stomach. After the young are born, the stomach resumes normal digestive
functions. In 1984 a second species of gastric brooding frog was discovered: the closely related northern gastric brooding frog, also from Queensland. Sadly, intensive searches for both species during the past decade have
turned up empty-handed. Gastric brooding frogs appear to be extinct.

POUCHES ON THE front, the back, along the side of the body, in the
vocal sacs, mouth, or stomach all provide protection. In the two invertebrates—the isopod and the starfish—the mothers stay put. In the vertebrate species, brooding the young in pouches allows the parent to go
about his or her activity while protecting the offspring.

And so it is with people. Human cultures the world over use makeshift
pouches, allowing us to gather food, plant crops, weave, herd livestock,
or surf the Internet knowing that baby is safe. In the Andes of Bolivia,
a toddler snoozes in a woolen shawl tied around his mother's shoulders
while she collects firewood. In the Canadian far north, a baby peers out
of the large bear fur hood on her mother's parka while they travel by dog
sled. In Papua New Guinea, an infant gurgles from a string bag hanging

from his mother's head while she prepares the evening meal. I suspect that nowadays in virtually every country one can find a mother surfing the Internet with her infant strapped to her chest in a baby carrier.

LEFT HOLDING THE EGG

Parents are the bones on which children sharpen their teeth.

PETER USTINOV, *Dear Me*

Any parent knows the sacrifices of time, energy, and financial resources that accompany the rewards of raising children. We humans aren't alone in parental sacrifice, however. Consider the emperor penguin, largest of the sixteen species of penguins and weighing up to ninety pounds. These birds not only breed under severe physical and environmental conditions; they also win an award for parental devotion.

Emperor penguins breed during the Antarctic winter in temperatures that sometimes drop as low as $-95°$F and in wind speeds that reach 120 miles per hour. Why do they chose this time to breed? The chicks need about a year of growth to survive the following winter on their own, so they must hatch in the dead of winter. If they hatched in the spring, fall, or summer, they wouldn't get in a full year's worth of growth before the next winter. The birds establish their rookeries (breeding grounds) on well-anchored sea ice, usually between the coast of Antarctica and small offshore islands. Often these rookeries are over seventy miles from the open sea, the penguins' only source of food.

Well before they establish their rookeries, during the austral summer (January through March) the penguins feed at sea and store body fat. By late March, ice covers the sea and traps the penguins' food. The rotund penguins begin their march toward the rookery, waddling along the ice often in single file at a rate of about 1.2 miles per hour—taking up to fifty hours to reach their destination. Once there, the birds court with loud trumpeting calls. Eventually they mate. Their chosen site doesn't seem a very hospitable place for birds to lay their eggs. No nesting material—just snow and ice.

During May each female lays a single white egg weighing about one pound. To keep it off the ice, the female rolls the egg up onto her feet and covers it with the lower fatty part of her belly. This area is a fold of loose skin that wraps the egg in a blanket, densely feathered on the outside and bare inside. The long fast and depletion of energy reserves spent marching to the rookery and producing the huge egg have used up to 30 percent of her body weight. She's hungry by this time, so she soon passes the egg over to the father's feet and shuffles back to the open sea to feed.

Like the mother, the father balances the egg on his feet and protects it with his fold of loose skin. As the weather deteriorates, he waddles, with the egg on his feet, into the growing group of other males left holding their eggs. In this position and in this location, the stay-at-home dads spend the next two months, huddled together in groups of hundreds or thousands of individuals. By huddling, each penguin has less surface exposed to the wind and cold.

As each day passes, the males become thinner and hungrier. Some lose as much as 50 percent of their body weight. By early July, the eggs begin to hatch. If an egg hatches and the mother hasn't returned yet, the father feeds his chick with a milklike substance from the lining of his esophagus. But he can do this for only so long. If his body weight drops to about forty-eight pounds and his mate still hasn't returned, he abandons his chick (or egg) and heads off to sea in search of food. If all goes well, though, the well-fed females return to the rookery before the chicks hatch. Each expectant male recognizes his arriving mate's call and he responds in kind, directing her to his location. After the male transfers the egg (or chick, if the egg has hatched) over to the mother's feet, he begins his long journey toward the open sea to replenish his own energy reserves.

Once back, the female feeds the chick by regurgitating undigested fish, squid, and crustaceans she has brought back in an internal pouch located off her esophagus. The male stays away for about four weeks, but being a responsible father, he returns to the rookery in August and relieves his mate. It's her turn to eat.

During September the chicks become less dependent on their parents, and they gather with others their own age. Finally, the parents can leave together to dine and to gather food for their increasingly ravenous chick. When each parent returns, he or she calls for its chick, and when the chick responds, it gets regurgitated food. Parents feed only their own chicks, because chicks respond only to their own parents' calls. As the ice breaks up, the parents' trips between the rookery and open sea become shorter, allowing them to gather more food to meet the increasing demands of their growing chicks. By December the ice beneath the breeding

grounds breaks up and the birds disperse. The chicks are now independent, after about five months of intensive care by devoted parents.

WHEN WE THINK of parents making supreme sacrifices for their young, we usually think of birds carrying insects and earthworms to their hungry chicks waiting impatiently back in the nest, and female mammals tirelessly nursing their young. But some fishes, amphibians, reptiles, and invertebrates make supreme sacrifices of time, energy, and risk to life and limb. For example, females of many true bugs, thrips, beetles, sawflies, and treehoppers, to name but a few, straddle their eggs and shield them from predators and parasites. By attacking intruders that would like to make a meal of their young, they put themselves in jeopardy of being eaten.

Death is the ultimate sacrifice for one's offspring. An attentive mother octopus watches over her many thousands of eggs until they hatch into miniature octopuses. She aerates the eggs by shooting water through her siphon, removes debris from the egg mass, and fights off potential egg eaters. She rarely eats while attending her eggs, and she will most likely die of starvation by the end of her vigil. Octopuses generally grow quickly, reproduce once, and die before they reach their third birthdays.

Here's a dramatic example of ultimate sacrifice: females of some mites that reproduce only once per lifetime can't lay eggs outside their bodies because the eggs are so large. Soon after forming eggs, the female dies, providing a safe haven in which the eggs develop. Eventually the young hatch inside their mother's corpse and make their way to the outside world.

My favorite is a pyemotid mite in which males usually complete their life cycle and die before they're even born! A typical batch of young consists of one male and about fourteen females. After hatching inside the dead mother's body, the male mates with his sisters and then dies. He never sees the light of day. His sole purpose in life is to mate with his sisters. The daughters make their way out of their mother's corpse and seek out grain beetle larvae for food. Once a female mite finds a larva, she inserts her mouthparts and sucks out liquid food—nourishment to support development of the eggs fertilized by her brother. Eventually she dies, her eggs continue to develop, and a new generation of mites hatches and mates within their dead mother's body.

Spiders also make supreme sacrifices for their young. In some species in which females guard or carry their eggs, the female lives to watch her young disperse out into the world.

In others, the female dies after caring for her young for a short time; after they hatch, the spiderlings dine on their deceased mother.

And then there's a remarkable Australian social spider. After the spiderlings hatch, the mother hunts and lugs back large insects for her young to eat. Some of these insects are up to ten times her body weight. Imagine hauling a cow ten times your weight through the grass to feed your babies! The female also dines on a fair share of the catch, storing body fat and developing additional eggs. With cooler weather, insect prey become harder to find, and both mother and young spiders have less to eat. Nutrients from the developing eggs seep into the mother's bloodstream. The hungry youngsters turn on their unresisting mother and suck the nutrient-rich blood from her leg joints until she is weak and withered. Once she's dry, they inject venom and eat the rest of her. They have attacked their first prey.

PARENTAL CARE OFTEN costs the parent considerable time and energy—time and energy that could be spent doing something else, like mating and making more babies. Parental care often exposes the adult to predators and may cause the parent to die sooner rather than later. But guaranteed death as a result of parental care really is the ultimate sacrifice. How could this behavior evolve? Short-lived species that can breed only once per lifetime have little to lose and everything to gain by giving their all to offspring.

Sets of reproductive behaviors, including parental care, are the result of natural selection. What works for individuals of a given species survives—until it doesn't work anymore. What doesn't work disappears over time. The parental behaviors we see today reflect one point in time. Behaviors, like other attributes of species, continually evolve with changing environments. The seemingly drastic parental sacrifices described here may disappear with time . . . or they may become even more drastic.

3 Eat to Live and Live to Eat

ANIMALS MUST EAT to live (and some of us seem to live to eat). Eating involves making decisions about what, when, and how often to eat, how to detect and capture food, where to search or wait for a meal, and when to leave an area to search or wait elsewhere. The choices an animal makes depend on whether its food is distributed evenly or in patches, food abundance, food quality, mobility of the food source, how many other animals are competing for the same resources, and which predators are lurking nearby ready to make them food. The sum total of these decisions is an animal's feeding behavior.

In our venture into unusual animal natural histories, we could find bizarre examples in any aspect of feeding behavior. We'll touch here on two areas: what and how.

First we'll look at some unusual foods: blood, dung, then some foods humans eat. We often think of other animals' diets as gross or unsavory at best, but humans eat their fair share of oddities as well. Whether or not you consider an item delicious or disgusting depends on your culture, your individual background, and your particular set of taste buds.

Second, we'll look at some unusual ways animals get their food, including gutless wonders that absorb nutrients from their hosts and some animals that lure prey by wriggling body appendages. Some animals work together to secure food, and they share the bounty with nest mates.

BLOOD MEALS: FROM BAT TO TICK ATTACKS

With that he pulled open his shirt, and with his long sharp nails opened a vein in his breast. When the blood began to spurt out, he took my hands in one of his, holding them tight, and with the other seized my neck and pressed my mouth to the wound, so that I must either suffocate or swallow some of the——Oh my God! my God! what have I done? What have I done to deserve such a fate, I who have tried to walk in meekness and righteousness all my days. God pity me!

BRAM STOKER, *Dracula*

Though I trust none of us has been transformed into a vampire, as was Mina by Count Dracula, we've all been victims of vampires—creatures that suck our blood. Think about it: we provide a critical link in the food chain by unwillingly offering our blood to flies, fleas, mites, ticks, leeches, bedbugs, and sometimes even bats. Mosquitoes and bloodsucking leeches are barely tolerable, but I draw the line at vampire bats. When I sleep outside on the ground in the tropics, I wear my shoes. I'm not offering up my toes to ravenous vampire bats!

The word "vampire" hails from eastern Europe, where it refers to a corpse that comes alive at night and sucks blood from sleeping people. The innocent victims wither away, die, and become vampires themselves. Vampire legends probably arose from the hundreds of savage murders committed in the 1400s by Prince Vlad Tepes, who lived near Transylvania, in what is now Romania near the Hungarian border. Prince Tepes gained fame and glory as a crusader against the Turks and was infamous for his gruesome actions. One of his favorite execution methods was driving sharpened stakes through his victims' bodies. The character of Count Dracula in Bram Stoker's horror novel is based on Prince Vlad Tepes. Dracula means "son of the devil" in Romanian.

Given that vampire legends were prevalent and widespread in Europe, Europeans naturally were horrified when the Spanish and Portuguese explorers of the 1500s sent back reports of blood-drinking bats from Central and South America. Appropriately, they called the animals vampire bats.

All three species of vampire bats occur only in the New World. Two feed on birds' blood. The third, the common vampire bat, feeds mainly on the blood of mammals—including humans. Common vampire bats are brown, with small pointed ears, no tail, and a wingspan of about fifteen inches. In many parts of the New World tropics, common vampire bats have become more abundant over the past few centuries, thanks to increased food: herds of cattle, horses, sheep, and other domesticated animals, plus growing human populations. In the 1500s indigenous peoples generally lived in small groups scattered throughout the landscape. Now villages and towns are common, offering vampire bats a veritable smorgasbord of skin.

How does a vampire bat obtain a meal without alerting its victim? For starters, it attacks only sleeping animals. The bat lands on or near its target. Weighing less than two ounces, the bat isn't likely to be noticed. It then crawls to a relatively bare patch of skin where copious blood awaits just under the surface: a dog's nose; a cow's neck; a pig's nose, ears, or teats; or a human's big toe. It sinks its razor-sharp teeth into the skin, scoops out a sliver of flesh, and inserts its tongue into the wound. Two channels on the underside of its tongue allow the bat to slurp blood as though it were drinking through a straw. The bite causes little pain, and the animal rarely awakens. Fortunately, vampire bats can't bite deeply enough to break any major blood vessels. The small, shallow wounds would normally stop bleeding within a minute or two, but a vampire bat needs more than one or two minutes to feed. Thanks to an anticoagulant in the bat's saliva, the blood keeps flowing and the vampire can feed for up to several hours from one wound while its victim sleeps soundly.

Feeding on blood isn't without potential problems. Vampire bats spend about twenty-two hours each day roosting in caves. They fly out at night, locate sources of blood, fill their bellies, and then return to their caves. Their stomachs can hold a volume of blood equal to 57 percent of their body mass, but they can't fly with this much extra weight. The problem is solved by rapidly getting rid of water and lightening their load before taking off. Within two minutes after a bat begins to feed, it begins to excrete a stream of very dilute urine. But having lost so much water, the bat has a new potential problem. Its belly is now full of high-protein food that will produce lots of urea. To excrete urea, most mammals need a lot of water to form urine. Vampire bats, however, have unusually flexible kidneys that can produce very concentrated urine. By alternating the production of very dilute urine (when feeding) and very concentrated urine (when resting and digesting), the bat can have its blood and drink it too.

HIGH IN PROTEIN, blood is the diet of choice for many animals other than vampire bats and Romanian princes. Blood is just another type of food. Bloodsuckers view the world from a different perspective than do predators, such as most humans. For a predator, the world is full of whole prey to be captured and eaten. For a bloodsucker, the world consists of mobile blood bars.

Most bloodsuckers use one of two techniques. One is to move around from host to host and tank up on the liquid food ("bar hopping"). The other strategy is to attach to a victim and draw frequent blood meals ("running up a tab"). Both appear to be successful ways of staying fat and happy, judging by the abundance of these animals. Let's look at some invertebrate blood feeders.

Bloodsucking leeches are renowned blood feeders: flattened and tapered worms ranging from less than an inch to eighteen-inch monsters from the Amazon Basin. Most live in freshwater ponds, lakes, and streams, though some live in oceans and others live in damp soil or on vegetation. Land leeches are fairly common in wet areas of Asia, Madagascar, and Australia, where they cling to damp leaves, poised with the mouth end in the air, waiting for unsuspecting victims. They cue in on vibrations, heat, and odor of mammals, including humans.

Using its posterior sucker, a leech firmly attaches to its host—usually a vertebrate. With its front sucker, the leech chemically digests an opening in the host's skin or slits the skin with its sharp mouthparts. Then the meal begins. These bloodsuckers have powerful muscles that quickly pump out blood. Leeches, like vampire bats, produce an anticoagulant (called hirudin in the case of leeches) that keeps the blood from thickening. In fact, hirudin is the strongest natural anticoagulant known. Most leeches gorge themselves and then drop off their hosts; they can survive months without another meal if it takes that long to find another host. Some freshwater leeches, however, live as external parasites attached to their hosts for long periods of time. When hunger strikes, they suck up a meal.

Leeches' bloodsucking abilities have long been appreciated in the medical field. For at least two thousand years, doctors and other caregivers used a European species now known as the medicinal leech (*Hirudo medicinalis*) to suck blood from patients suffering from mental illness, headache, rheumatism, gout, nosebleed, whooping cough, sore throat, skin disease, tumors, asthma, obesity, and drunkenness. Blood was removed because it was thought to be "in excess" and responsible for the patient's illness. Have you ever been told when you donated blood that getting rid of all that fluid was "good for your health"? It's the same misguided principle—today's version. Medicinal leeches are still used today, but for more sound reasons: to reduce accumulation of blood in tissues following plastic surgery, and for removing excess blood from severed ears or fingertips that have been surgically reattached. Furthermore, because the leeches secrete chemicals in their saliva that prevent clotting, the danger of excessive swelling in the tissues is reduced. A medicinal leech can consume up to five times its weight in blood. Then it takes an extended vacation for several months.

Before I left for my first field trip to the rain forest of eastern Ecuador, my dad told me he hoped I wouldn't have to deal with bloodsucking leeches. I asked if he meant leeches like the ones that attacked Humphrey Bogart while he was trying to extract the *African Queen* from the mud. Bogart's memorable line was, "If there's anything I hate in the world, it's leeches— filthy little devils."

Dad laughed, said Bogart's leeches were nothing, and relayed his experiences with bloodsucking land leeches when he was a guerrilla on Mindanao Island in the Philippines during World War II. When he slept on the ground, he would awaken with leeches dotting his body. No body part was off bounds when they wriggled underneath clothes to reach his skin. While hiking through dense jungle, he was constantly aware of the need to brush leeches off before they fastened on and caused festering sores that took forever to heal in the heat and high humidity.

Bloodsucking land leeches are also a nuisance in Australia and Tasmania. My friend Kay relates the following story. While spending part of a sabbatical year in Tasmania, one night Kay looked across the dinner table at her field biologist husband (a well-known ecologist who will remain unnamed to preserve his reputation) and was horrified to see an engorged black leech fall out of his nose and land in his mashed potatoes.

Another familiar blood drinker is the tick. Ticks (relatives of spiders) are ectoparasites. These eight-legged creatures attach to the outside surface of humans and other animals, puncture their host's skin with their beaks, and draw out blood. The more they drink, the more their bodies swell. Although many ticks have specific common names such as dog tick, chicken tick, sheep tick, deer tick, or cattle tick, most attach to other hosts as well. Some kinds of ticks inject toxins into their hosts and occasionally cause paralysis in people they've bitten. Word of advice: If a tick embeds itself into your skin, the best way to remove it is to grasp the mouthparts as close to the skin as possible with tweezers and slowly but firmly pull. Don't grab the tick by its body because the body may break off, leaving the head or mouthparts in your skin and causing infection.

Many kinds of flies feed on vertebrate blood. We fall victim to many of these vampires. In some species—such as tsetse flies, stable flies, and horn flies—both males and females feed on blood. In others—such as mosquitoes, blackflies, biting midges, sand flies, and deer- and horseflies— only the female drinks blood. After mating, female mosquitoes seek out their victims. In some species, a female needs the blood's amino acids to

begin developing her eggs. In other mosquitoes,
the blood increases the rate of egg production.
Generally, the more blood a female ingests, the
more eggs she'll produce. Tabanids (horseflies
and deerflies) have large heads, large eyes,
and especially large and sharp piercing
mouthparts—in some African species, strong enough to pierce through
tough hippopotamus hide. In areas where tabanids are especially dense,
they may consume as much as one-sixth of a pint of blood per day from
a grazing animal.

Back when the fashion was to dock the tails of carriage horses, the
French charged lower rates for boarding horses with docked tails than for
animals with unadulterated tails. Why? Horses are so pestered by flies that
without a fly-swatter tail, a horse eats less. Eat less, you pay less.

Another insect vampire, the bedbug, feeds on human blood. By day
these loathsome insects hide out in cracks and crevices of a mattress or
under a rug. When night falls, they stealthily emerge. They puncture the
skin of a sleeping victim, pump saliva into the wound, and then suck up
the mixture of their own saliva and the host's blood. The saliva is probably
responsible for the resulting itchiness and the victim's sleepless night
thereafter. Bedbugs can live for several months without feeding, patiently
waiting inside the mattress for the budget hotel's next guests.

EVEN SOME HUMANS other than Romanian princes consider blood
just another source of nutrition. Have you ever eaten blood sausage, blood
cheese, or blood soup? Many people consider these foods delicious. Hunters often bleed the rabbits or wild birds they've killed, saving the blood to
thicken gravy that they'll slather over the roasted meat. Unlike other blood
feeders that "bar hop" or "run up a tab," though, humans usually kill their
victims before they collect the blood. From a mosquito's, bedbug's, or vampire bat's point of view, that wouldn't make much sense—blowing up the
beer kegs, so to speak!

EARTH'S SANITIZERS

Egyptians believed that the activities of the dung scarab represented their own
world in miniature. Every year these industrious creatures buried balls of dung,
and every year more beetles suddenly emerged from the ground. In ancient Egypt
the sun god, Ra, was symbolized as a great scarab, rolling the sun, like a dung

ball, across the heavens of a universe with Earth at its center. The burying of the
dung ball came to symbolize the rising and setting of the sun.

ARTHUR V. EVANS and CHARLES L. BELLAMY,
An Inordinate Fondness for Beetles

Coprophagy, the eating of feces, occurs in diverse animals—both invertebrates and vertebrates. Excrement is just another type of food for some
animals. Let's look at a sampling.

Green iguanas rely on a microbial fermentation system in their hind-
guts for up to 30 percent of their energy requirements. Baby iguanas need
these microbes (bacteria and protozoans) to break down the cell walls of
the plants they eat, but they don't have the microbes when they hatch from
their eggs. These herbivorous lizards get their hindgut microbes by eating
fresh feces from adult iguanas. Because there aren't any adult iguanas near
the nests, the young must seek out their elders.

During their first week after hatching, the babies eat soil from the nest
chamber and from around the nest. Soil bacteria provide a simple fermentation system and help the hatchlings digest vegetation during their second week. But it's not enough. At that point, the young disperse into the
forest canopy where they seek out adults—and their feces. After they've
gotten inoculated with the critical microbes a few weeks later, the young
climb back out of the canopy and migrate to forest-edge vegetation, where
they live for several years. Once they reach adult size, they return to the
canopy.

Rabbits and their relatives have an enlarged sac called a cecum, in
which bacterial fermentation takes place. Rabbits produce two types of
feces: a daytime variety and a nighttime one. The nighttime feces are chock-
full of bacteria. Rabbits eat the nighttime feces, recycling the bacteria and
absorbing additional nutrients in the process.

One unusual invertebrate dung-eater is the larval stage of the sloth
moth, *Cryptoses choloepi*, which feeds on the fecal pellets of three-toed
sloths. Adult female moths live in the thick fur of sloths, safe from bird
predators. The sloths descend from the trees to the ground to defecate
once a week. Sloths dig depressions with their hind claws, deposit about
a cupful of pellets into their "toilets," and then cover their jobs loosely
with leaf litter. Female moths ready to lay their eggs leave the sloths at this
point and deposit their eggs in the fresh dung. After hatching, the larvae
feed on the dung and eventually pupate. Once the moths emerge, they fly
into the trees, find sloth shelters, mate, and carry on the life cycle.

AND THEN THERE are dung beetles, feces-eaters extraordinaire. The sacred dung scarab of Egypt symbolized resurrection and immortality, explaining why the ancient Egyptians often placed these beetles in tombs with the dead and why they painted dung beetle figures on their sarcophagi. During the embalming process, they sometimes removed the deceased's heart and replaced it with a large scarab charm carved from green feldspar or obsidian. Scarab charms were thought to bring the bearer good health and happiness in death as well as in life.

Dung beetles haven't always been universally admired, however. The dung beetle symbolized sin in medieval Christianity. Many people today see dung beetles as disgusting creatures because they crawl around in and eat manure. These beetles do us a tremendous service, though. They bury dung—not to clean the ground for us, but as a source of food for their larvae. Without these sanitizers, we'd literally be in "deep shit." For example, during parts of the year in India, dung beetles bury an estimated forty to fifty thousand tons of human excrement each day.

Some species of dung beetles aren't particular about the kind of excrement they eat, while others specialize on dung from only one type of animal. As a group, the two thousand or so species of dung beetles eat just about every kind of dung there is, from delicate bird droppings to enormous piles of elephant manure. Several species of rain forest dung beetles collect monkey dung from high in the trees and then tumble to the ground clutching their treasures. Some species cling to the fur of their preferred dung source—monkeys, sloths, wallabies—and get their food hot off the production line. Dung piles can attract huge numbers of beetles. In Africa a mound of fresh elephant dung can attract four thousand dung beetles within fifteen minutes.

Different species use the dung differently, but basically dung beetles eat it and lay their eggs in it. When the eggs hatch, the larvae have a bountiful source of nutritious food. Some species lay their eggs in the mother lode itself. Others carve or scoop out chunks of dung, which they bury under or near the source. Some industrious species build complex underground nests with tunnels, into which they stuff their dung chunks. Some form dung into balls and then roll the balls a long way, often more than a hundred yards from the source.

Why do some species roll the dung so far? Dung piles are widely dispersed and they don't stay fresh very long. For these reasons, they're the objects of considerable competition. As dung dries out, the beetles can't cut or chew the solid mass. Everybody and his beetle brother is out there looking for fresh dung, so when a beetle finds it, it's best to carve out a share and run with it. Once the beetle has his ball, he sometimes must defend it against unscrupulous individuals that attempt to steal his hard-earned treasure. To protect his prize, a male will climb onto his dung ball, hold on with his four back legs, and use his front two legs to repel attackers, flipping them off as they attempt to scale his booty.

The sacred scarab from Egypt and the Mediterranean region is an example of a dung ball roller. Upon finding a fresh cow pie, a male plunges in and begins scooping and digging. Using his front pair of legs, he pushes a pea-sized wad between his second and third pair of legs. Holding the pea with the claws of his hind legs, he spins the dung between them. In time, the sphere becomes the size of a marble, then a golf ball, and eventually an apple. Once satisfied with its size, the little beetle (who himself is only one and a half inches long) rolls the ball away from the source. Gripping the ball with his back pair of legs, he stands on his front legs with his head facing the ground. By moving his back legs, he sets the ball in motion. With his front legs walking backward, he continues until he finds a suitable final resting spot. There a female will eventually excavate a large brood chamber several inches below the soil surface and lay eggs in her mate's treasure.

How important are dung beetles to humans? Consider Australia, where the indigenous dung beetles use the dry feces of native marsupials such as wallabies and kangaroos. When cattle were introduced to Australia, the local beetles ignored the dung. They'd never been exposed to anything as soggy as cow pies. In many parts of the world, a cow pie lasts only a few hours to a few weeks thanks to dung beetles. In Australia, however, a cow pie remained undisturbed for up to five years. An average cow produces ten to twelve pies each day. Multiply this by a large herd of cattle and many large herds across the continent, and you've got a problem—an estimated 200,000 cow pies per minute problem!

With no cow pie dung beetles in the land, Australia faced two problems. First, cow pies serve as breeding grounds for flies that threaten human health. Second, extensive pastureland was lost because the cows (understandably) considered grass growing next to their pies to be less than tasty.

Health concerns and economics provided incentives to remove the accumulating dung, and in 1963 Australian authorities decided to import foreign dung beetles. Because the savannas of southern Africa are similar in climate to the cattle-raising areas of Australia, entomologists chose four species of African dung beetles. To avoid introducing unwanted microorganisms such as mites, roundworms, fungi, and bacteria normally found on the beetles, the biologists imported eggs rather than adults.

Technicians extracted eggs from dung in Africa and washed the eggs before shipping them to Australia. There workers transferred the eggs to cattle dung and allowed them to hatch into larvae, pupate, and emerge as adults under controlled conditions. The adults mated and laid eggs, creating a second generation of eggs that served as the foundation of a mass rearing program. Finally, in 1967 Australians released adults of these four species of dung beetles with the fervent hope they would do their dung duty. One species quickly established itself, and Australia's cattle dung problem has been alleviated. (This is one of the few successful animal introductions to improve human "quality of life." Most haven't worked.)

DUNG BEETLES COMPETE with other insects, such as horn flies, for excrement. Female horn flies lay their eggs in cow pies, but first they must feed on cattle blood for protein to develop their eggs. Sometimes horn flies become so dense—up to several thousand flies per cow— that they greatly reduce weight gain of the cattle. Needless to say, ranchers despise horn flies. After mating on a cow, the female flies off the animal, lays her eggs on or underneath the fresh dung, and within five minutes is back on the cow feeding on more blood. But if dung beetles are around, the fly larvae that hatch from those eggs won't have the dung pile to themselves. Dung beetles break up the cow pies and destroy these food resources for the flies. Ranchers appreciate the beetles because they provide a great natural control for these pestiferous flies.

Dung beetles are worth their weight in gold—or at least in green feldspar and obsidian. The next time you hear about or see a dung beetle, don't be repulsed. Think about the service it's doing, sanitizing the earth as it devours its favorite food and buries chunks of nutrients for its larvae.

ONE PERSON'S PLEASURE IS ANOTHER PERSON'S POISON

Grasshopper Tacos

$\frac{1}{2}$ lb. grasshoppers
2 cloves garlic, minced
1 lemon
salt
2 ripe avocados, mashed
6 tortillas (corn or flour)

Roast medium-sized grasshoppers for 10 minutes in 350° oven. Toss with garlic, juice from 1 lemon, and salt to taste. Spread mashed avocado on tortilla. Sprinkle on grasshoppers, to taste.

PETER MENZEL and FAITH D'ALUISIO, *Man Eating Bugs*

There's just no accounting for some folks' tastes. Hyenas, vultures, hagfish, starfish, snails, and flies are fond of rotting flesh. The eight-legged house-dust mite lives on dandruff. Whale lice feed on whales' dead skin. A hungry cockroach will nibble on fingernails of a sleeping person. Hovering around the eyes, ears, nose, lips, and udders of cows, flies slurp up tears, blood, mucus, saliva, and milk. Beetles, fly larvae, and crows feast on dung. Rabbits and guinea pigs eat their own feces. Vampire bats, leeches, bedbugs, kissing bugs, fleas, and ticks specialize on blood.

Some may think of rotting flesh, dandruff, dead skin, fingernails, body secretions, dung, and blood as odd gastronomic items, but the human diet also includes some unusual foods. Depending on what part of the world you call home, you may begin your meal with a soup of kangaroo tail, shark's fin, chicken feet, or blood. As a second course, you may choose boiled reindeer tongue, baked prairie dog, aardvark steak, hippo burgers, field mice casserole, roasted howler monkey, blood stew of beef entrails, or water beetles marinated in ginger and soy sauce. If you live near the ocean, you may eat sea cucumbers, sea squirts, sea urchins, and sea slugs. You may consider snake gallbladders, octopus tentacles, deer penises, pig ears, elephant trunks, ox tendons, pig brains, or marrow of giraffe bone a delicacy. Or what about live termites, fried caterpillars, chocolate-covered ants, or pickled beetle pupae? For fat you may eat whale blubber, camel hump, hippopotamus lard, cockroaches, or beetle grubs.

We're obsessed with food and with good reason. We can't live without it. It also adds pleasure to life. After a difficult or traumatic experience, we crave comfort foods: chocolate chip cookies, homemade chicken soup, macaroni and cheese. Our social interactions and cultural rituals center on food: potluck dinners, banquets, neighborhood barbecues. We eat food raw, roasted over an open flame, baked in a clay oven, cooked on an electric or gas range, or warmed in a microwave. Some prefer food au naturale. Others drench their food with spices to change or improve the flavor.

You've probably compared notes with others on the odd food you've eaten. "Have you ever tried roasted caterpillars? Or fried cat?" How we define odd, though, depends on our culture. For example, most North Americans and Europeans adore cheese. In contrast, Chinese visitors to our country often avoid cheese. They wonder how we could possibly eat, much less like, rotten milk.

Join me on a trip around the world to sample some unusual cuisine.

HUMANS ARE THE only animals (except for flies) to drink milk meant for the young of other animals. Not only that, but milk is considered one of the most nourishing foods (except for those who are lactose-intolerant). Depending on where you live, you might drink milk from cows, sheep, goats, water buffalo, yaks, reindeer, camels, llamas, horses, or even zebras.

Some adults leave milk to the kids, preferring alcoholic drinks made from rotting fruits (wine), grains (beer, whiskey, sake), stalks (rum), and roots (yuca *chicha*). Fermentation happens when bacteria, mold, and yeast break down sugar. People worldwide make fermented drinks, and it's not a recent invention. Over five thousand years ago, residents along the Nile River brewed beer. Ancient Egyptians and Babylonians grew grapes and crushed them into wine. And we all know how much the ancient Greeks and Romans loved their wine.

South Americans make *chicha* from yuca (also known as cassava or manioc), the thick, starchy root of a small shrub. I was introduced to this fermented drink while I was a guest at a Quechua wedding celebration in eastern Ecuador. Sitting in a circle on the bamboo floor of a thatched hut, eight Quechua women talked and laughed as they chewed chunks of yuca. Every few minutes they spat the accumulated saliva and wads of chewed fiber into a large hollowed-out gourd. The mixture would be set aside for a few days to ferment and then strained. Enzymes in the women's saliva helped to speed the fermentation process.

Meanwhile, a batch of *chicha* prepared several days earlier made its rounds of the guests. I felt queasy as the bride's mother dipped a small gourd into a larger container and offered me the yellow fermented broth. Somehow I swallowed the strong and bitter brew. Although the *chicha* wasn't slimy as I had expected, I kept thinking about the saliva. The bride's mother came around often. I soon faked the sips.

Tuba, or coconut sap, is a popular drink in the Philippines. People collect sap from coconut palms and then strain it through coconut fiber to remove the drowned insects. If the *tuba* is to be fermented, it's filtered through bark, which gives it a slightly bitter taste. The insect-free sap is poured into a coconut husk and left in the sun to ferment. While my dad was on Mindanao during World War II, he drank *tuba* juice for breakfast whenever he could get it from the locals. After Dad returned to the States, a doctor told him he could credit *tuba* for sparing him any number of nutritional ailments, thanks to the high vitamin and mineral content of coconut sap.

Filipinos strain insects from *tuba,* but Mexicans add caterpillars of the giant skipper butterfly to bottles of *mezcal.* (The caterpillar is sometimes called a "worm" because of its wormlike body shape.) Caterpillars are added because their presence certifies authenticity. Real *mezcal* is produced only from the agave plant. That's the only place you'll find the skipper caterpillars, because that's all they eat. Presence of a well-preserved caterpillar also verifies that the brew has not been unscrupulously watered down. If it had been watered down, the larva would have rotted.

The Chinese add animals to fermented drinks to increase potency and medicinal benefits. Wu Shiu Jiu (five-snake wine) is a medicinal drink thought to invigorate the body and help with joint ailments. Five different kinds of venomous snakes are eviscerated, dried in the sun, and then soaked in a large jar of distilled rice wine. Another Cantonese specialty is gallbladder removed from a live snake and dropped into a glass of wine, drunk in the belief that one will take on the snake's strength and be protected from sickness. Ant-steeped rice brandy, a popular drink in China, is believed to be effective against hepatitis B and rheumatism thanks to the formic acid and minerals from the ants.

WHAT GOES BETTER with Cabernet Sauvignon than a tender, juicy filet mignon? You and I might choose that for our animal protein, but some people prefer animals other than cattle. Animal dishes offer some great examples of unusual human diets.

Take, for example, animals' stomachs filled with stuffing. Haggis, a national dish of Scotland, is a sheep's stomach filled with sheep liver, heart, and lungs mixed with animal fat, onions, oatmeal, and seasonings, boiled to perfection. We have a comparable dish in the United States: roasted pig's stomach, made by filling a pig's stomach with ground sausage, potato, onion, cabbage, and spices. The stomach is sewn closed and then roasted until golden brown and crispy.

Camp cooks in the American West invented the appetizer known as Rocky Mountain oysters. During roundup time, after the bull calves were castrated, the cooks fried the testicles in batter and served them up as a treat for the hardworking cowboys. A traditional Basque recipe calls for simmering the testicles with garlic, onions, red peppers, seasonings, and white wine.

Recently a friend told me about "head cheese." As a little girl in Iowa, Kay often enjoyed eating head cheese sandwiches at her grandmother's house. Not a cheese at all, head cheese is bits of shredded meat in a jelly matrix. One day Kay saw a large box on her grandmother's kitchen counter. As she started to open the box, her grandmother admonished, "Don't open that!" Kay opened it anyway—and found a pig's head staring up at her. Her grandmother's response? "I'm going to make head cheese today."

Head cheese can be made either from the head of a pig or calf. The recipe from *Joy of Cooking* calls for a calf's head. After quartering the head, you're instructed to clean the teeth with a stiff brush. Then remove the ears, brains, eyeballs, snout, and most of the fat. Set the brains aside. Soak the quarters in cold water for six hours to remove the blood. Then simmer the quarters two or three hours with onions and celery sticks. Drain and set aside the stock. Remove the meat from the bones. Dice the meat and then cover with the stock. Add salt, pepper, and herbs. Cook thirty minutes. Pour into a mold and chill. Cut into slices and serve with a vinaigrette sauce to which you've added the cooked brains.

Meat from just about any animal is eaten by someone, somewhere. Acquired tastes in the United States include skunk stew, opossum baked with sweet potatoes, porcupine liver, poached alligator tail, baked rattlesnake, and bullfrog chop suey. Yak tail is a highly esteemed food in Tibet, and electric eel is relished in South America. The Cantonese commonly eat snake: deep-fried snake meatballs, fried shredded snake with vegetables, and soup made—once again—from five kinds of snakes. Prairie dogs are eaten in the United States, guinea pigs in South America, hamsters in China, and hedgehogs in Britain. The original Australians and early settlers to Australia ate platypus. Flamingo tongues were considered gourmet fare in ancient Rome and were often served at emperors' banquets.

"Tiger Fights Dragon" is a popular dish in Canton—a combination of cat meat and snake meat. The waiter comes to your table with a live snake and asks if you want the gallbladder added to your meal. If yes, he cuts the snake open on the spot and removes the quivering organ. The gallbladder, believed to provide health benefits, will cost you extra.

Dog meat is unthinkable to most North Americans, but it's widely eaten in China. There people raise dogs for food just as we raise rabbits and South Americans raise guinea pigs. Dog meat is especially popular during the cold winter months because it's believed to generate extra body heat.

Some people prefer fish to red meat. And not just any fish. *Surstromming,* sour Baltic herring, is a Swedish delicacy. Fishermen immerse the herring in brine for a day, then decapitate and clean them. Afterward they pack the fish in barrels and leave them in the sun for twenty-four hours to initiate fermentation. Wisely, they leave an inch or two of space at the top of each barrel so that accumulating gases won't explode the barrel. Then they store the barrels in a cool place and let the fish ferment at a slower rate for up to a year. I'm told that despite the ripe smell and sharp taste, the fish is delicious.

Raw fish is commonly eaten in Japan as sashimi (thin slices of raw fish) and sushi (rice, raw fish, and vegetables wrapped in seaweed). Many North Americans once considered it uncivilized to eat raw fish. No longer squeamish, these same people now consider raw fish a delight. Our tastes change. Sashimi and sushi bars are now popular all over the United States.

Insects and spiders are other unusual sources of animal protein. Many groups of Indians in the Amazon Basin and some in Central America eat leaf-cutter ants, either raw or roasted. Young Bedouin boys in the Egyptian deserts eat dung beetles as a coming-of-age rite of passage. People from various tribes in Africa eat caterpillars, beetle larvae, crickets, and termites. Cambodians eat cicadas either raw or impaled on a stick and roasted over an open fire like frankfurters. People in China, Australia, and South America eat cockroaches; people in tropical Africa intentionally eat mosquitoes (as opposed to those that simply fly into a hiker's mouth); and select groups of people around the world enjoy fly maggots.

Authors of the book *Man Eating Bugs,* Peter Menzel and Faith D'Aluisio traveled widely, sampling insect and spider culinary delights. Some of their discoveries include scorpion soup (China), dewinged dragonflies fried in coconut oil (Indonesia), stir-fried giant red ants (Thailand), grub dip (Australia), sun-dried worms (Botswana), palm grubs sautéed in their own oil (Uganda), mealworm spaghetti (Mexico), roasted tarantula (Venezuela), and "Cricket Lick-It" crème de menthe–flavored lollipops (United States).

How do insects and spiders taste? Menzel describes for us his experience eating roasted giant tarantula in Venezuela: "When we crack them open, there's white meat. No goo—this creature has actual muscles. The same with the abdomen, which has the most meat. It's tasty—like smoky crab. I wish I had a half-dozen more." To Menzel, Chinese stir-fried cicadas have a "crispy mild taste like salty bacon puffs"; Peruvian grubs taste like "pork sausages with crunchy heads." He describes deep-fried silkworm pupae from China as follows:  "Each one pops in my mouth when I bite down, re- leasing a rush of flavor not unlike what I imagine a deep-fried peanut skin filled with mild, woody foie gras would taste like." Not all the dishes he tried were pleasant to his palate, however. He describes stinkbugs eaten in Mexico as "an aspirin saturated in cod liver oil with dangerous sub- currents of rubbing alcohol and iodine."

In her fascinating book *America's First Cuisines,* Sophie Coe describes the foods eaten by the Incas, Maya, and Aztecs. In addition to domesti- cated animals and wild game, these early inhabitants of the New World re- lied on insects for animal protein. The Incas relished mayfly larvae, which they either ate raw and alive or as sauces of toasted larvae combined with chili peppers. They also ate ants, caterpillars, and beetles. The Maya ex- tracted large beetle grubs from rotting trees, toasted them with salt, and then wrapped the grubs in tortillas. They roasted skewered dragonflies over open fires. Aztecs collected water boatmen (a type of water bug) in nets, ground them, rolled small masses into balls, and then wrapped the balls in maize husks before cooking them. They roasted and salted cater- pillars—the same skipper caterpillars that are now added to bottles of *mezcal.* If you wish to experience some of the Aztec insect delicacies, visit Don Chon, an upscale restaurant in Mexico City. There you can order and enjoy toasted crickets, ant eggs sautéed in butter, and fried fly maggots.

It makes sense to eat insects. Consider grasshoppers. David Madsen and other archaeologists discovered that hunter-gatherers who lived near the Great Salt Lake in Utah five thousand years ago ate lots of grass- hoppers. Nutritionally, grasshoppers are great. Dried grasshoppers are 60 percent protein and 2 percent fat. A pound of grasshoppers yields about 1,365 kilocalories. Like other animals, people should rely most heavily on food resources they can gather most efficiently. How easy is it to collect grasshoppers? Madsen and his colleagues explored the econom- ics of grasshopper-eating. They collected sun-dried grasshoppers from the beaches around Great Salt Lake and found that one person could collect an average of two hundred pounds of hoppers per hour—about

273,000 kilocalories. To be conservative (there aren't always so many sun-dried grasshoppers available around the lake for the picking), they used one-tenth this figure to compare with the caloric yield of collecting sun-flower seeds and pine nuts and of hunting deer and antelope. The very conservative estimate of 27,300 kilocalories per hour compares favorably. A person can gather only between 300 and 1,000 kilocalories of seeds from around the lake per hour. Hunting deer and antelope yields about 25,000 kilocalories (twenty pounds of meat) per hour.

The nutritional and economic benefits likely explain why, in more re-cent times, Native Americans from the western United States often ate grasshoppers. Some roasted the insects and ate them as you would eat beer nuts. Some roasted and ground the grasshoppers and then mixed them with pine nuts, grass seeds, and berries to make cakes. After drying the cakes in the sun, they could store their prepared snacks for a long time.

NOT EVERYONE HAS stinkbugs, grasshoppers, yaks, or cattle avail-able. Some people simply don't like or can't eat meat. For those, cheese is a popular substitute. And we do make some strange cheeses. Limburger cheese, with its eau de dirty socks, must truly be one of the more raunchy-smelling foods some of us eat. We inject mold into cheese and *voilà!* Roque-fort and blue cheese. Some of the same people who throw away bread with the teeniest speck of blue-green mold consider Roquefort exquisite. Tibet-ans make cheese from yak milk; Laplanders make it from reindeer milk. Zebra, water buffalo, and camel milk also make edible cheese.

Eggs stolen from chickens, ducks, and other birds are another good meat substitute. If you have the opportunity, try a thousand-year-old egg, the unusual appetizer invented by the Chinese. The cook preserves raw duck eggs by coating them with a mixture of ashes, lime, salt, and tea, then leaves them to cure for up to six months. The chemicals soak through the eggshells and stain the insides blue and green. Eventually the contents become firm, smooth, and creamy. The taste is faintly fishy or cheeselike, not eggy.

And of course we eat more than birds' eggs. Consider the extrava-gance of caviar: the salted roe (eggs) of fish. The finest caviar is made by the Russians from sturgeon eggs. Sturgeon are huge primitive fish with long snouts, toothless mouths, and sucking lips. Their slender bodies and heads are covered with bony plates. A single Russian sturgeon may yield over four hundred pounds of caviar from a body that can reach twenty-five feet in length and weigh twenty-five hundred pounds. Shad, cod, salmon, tuna, and gray mullet eggs make less expensive caviar substitutes.

Central Americans eat the rich eggs of green iguanas and spiny-tailed iguanas. Finding a burrow full of already-laid iguana eggs is fine, but more prized is capturing a pregnant female iguana, because eggs still in the female's body are believed to increase one's sexual potency. A female stewed with her eggs is said to be "sublime."

Sea turtle eggs provide an important source of protein for people in some parts of the world, especially during the monsoon season, when fishing is difficult and unproductive. In other places, sea turtle eggs are gastronomic treats served as appetizers with cold beer, or they are highly prized aphrodisiacs.

People also eat invertebrates' eggs. In China, cooks use yellow crab eggs in asparagus and Chinese cabbage recipes. Sea urchin eggs are prized worldwide. Water boatman eggs are eaten in Mexico, ant eggs in Laos, and snail eggs in Paris.

ALL OF THIS was to convince you that other animals' diets are no odder than ours. The next time you're tired of raw oysters or pickled pig's feet, and your supermarket is just out of cat eyes, stinkbug pâte, or dried earthworms, be adventurous—try the recipe for grasshopper tacos.

THE INCREDIBLE EDIBLE EGG

Nothing stimulates the practiced cook's imagination like an egg.

IRMA S. ROMBAUER and MARION ROMBAUER BECKER,
Joy of Cooking

All higher animals pass through an egg stage. Without eggs, there'd be no offspring. Or parents. After all, which came first, the chicken (earthworm, snail, spider, butterfly, fish, frog, alligator, or mockingbird) or the egg?

Eggs also sustain life when they are eaten by other animals. Weasels, skunks, ferrets, and minks raid the henhouse. Flying squirrels, raccoons, opossums, and some snakes steal from bird nests. Mongooses and some monitor lizards tear open crocodile nests to get to the incubating eggs. Fish, salamanders, snakes, and turtles eat frog eggs. Crabs, birds, raccoons, pigs, and some monitor lizards and snakes dig up turtle nests and devour their eggs. Some frogs eat frog eggs, some fish eat fish eggs, and a few birds eat bird eggs.

Why are eggs so sought after? For starters, they're nutritious and full of energy. After all, eggs that develop outside the mother's body contain all the nourishment the developing embryo needs until it hatches. A chicken's egg is high in protein, iron, phosphorus, vitamins A and D, and most of the B vitamins. It contains about 75 kilocalories, of which 61 are in the yolk—the embryo's main food source. Second, eggs are often abundant. A bullfrog may lay twenty thousand eggs at a time. That's a lot of food for a hungry fish. Third, eggs are often easy to find. Gulls and many other seabirds lay their eggs in the open on beaches or rocky cliffs, making them prime targets for birds and mammals, including humans, who have probably stolen birds' eggs for food ever since prehistoric times. Fourth, eggs are easy to capture because they can't escape—although a predator often must deal with a protective parent defending its offspring. Let's look at some ways animals get to the insides of eggs.

BECAUSE FISH AND amphibian eggs are protected only by thin membranes, they're relatively easy for predators to handle. Many animals that feed on these eggs swallow them whole. One frog that eats frog eggs whole is the coquí, the national frog of Puerto Rico.

The Puerto Rican coquí, a golden-brown two-inch frog, lays large eggs (2.8 millimeters) on land, typically inside a curled leaf or palm frond. The eggs hatch in seventeen to twenty-six days, and tiny frogs rather than tadpoles emerge. Little calorie bombs, these large eggs contain enough energy to see the developing embryos through to hatching. An average egg contains nearly 52 calories (0.052 kilocalories), as compared to 2 calories per small egg (1 to 1.4 millimeters) of certain treefrogs that lay their eggs in water.

An average coquí clutch consists of twenty-eight energy-rich eggs, so a predator discovering a nest finds about 1,440 calories. Occasionally invertebrates such as humpbacked flies, ants, centipedes, millipedes, spiders, and snails chew into and partially consume eggs. The main predators, though, are intruding male coquís, which engulf the eggs whole. One egg clutch provides a male coquí with the equivalent of ten days' worth of food. If you ate ten times your daily food intake (20,000 kilocalories instead of 2,000), you could have the following: 20 eggs scrambled in butter, 15 strips of bacon, and 10 pancakes for breakfast; 12 pork link sausages in buns, 10 cups of fruit cocktail, and 3 pieces of chocolate cake for lunch; 15 small lobsters broiled in butter, 15 baked potatoes, 20 ears of corn, 51 carrots, and 5 pieces of cheesecake for dinner. Clearly, a male coquí feasts when he devours all the eggs in another male's nest.

It's not easy to do this, however, because male coquís guard their eggs. When an intruding coquí appears, the resident calls aggressively, and if the intruder doesn't leave, a battle involving biting, wrestling, and chasing ensues. It's no wonder males try to cannibalize other males' eggs: they feast and at the same time reduce the other male's reproductive effort. It's also no wonder that males guard their valuable eggs.

Certain tadpoles also swallow frog eggs whole. And some don't even have to search for their dinners. Conveniently, their mothers lay eggs for them to eat. This unusual behavior occurs in at least six species of New World tropical poison dart frogs, three species of New World tropical treefrogs, and one Asian treefrog. The commonality among these disparate frogs is that their tadpoles develop in small isolated accumulations of water such as bromeliad tanks, between leaf axils of plants, or in tree holes. These isolated sites may be relatively safe from tadpole predators, but they don't provide much food for the tadpoles. The solution? As mentioned in chapter 2, the mother delivers food: her own eggs.

Mothers of these two groups of egg-eating tadpoles lay their eggs in contrasting sites. The poison dart frogs lay their eggs on land. One of the parents guards the fertilized eggs and may keep them moist by urinating on them. After the eggs hatch, the parent carries the tadpoles piggyback to the water-filled tank in a bromeliad or some other small accumulation of water. In contrast, the four species of treefrogs lay their eggs directly in the water of bromeliads or tree holes. After hatching, the tadpoles develop there.

Periodically, the females of all ten species return to their tadpoles and lay additional eggs in the water. In some species, the male accompanies the female, and she lays fertilized eggs; eggs that aren't eaten will hatch. In other species, the female goes alone and deposits unfertilized eggs. Without the eggs, tadpoles of many of these species would die.

REPTILES AND BIRDS have evolved protective coverings for their eggs. And because these eggs are protected, predators have evolved numerous ways of gaining access to the delicacies.

Tough leathery shells protect lizard eggs from some predators, but other egg-eaters can extract the contents. Scarlet snakes and leaf-nosed snakes have a pair of saberlike teeth on the back of the upper jaw that slice through lizard eggshells. Although most land snails eat rotting plants and other detritus, they have teeth and some eat lizard eggs. Fire ants also chew into the leathery shells and feast on the oozing liquid.

The brittle shells protecting bird eggs aren't the most digestible (or delectable) parts of an egg. Nonetheless, many snakes swallow bird eggs whole—shells and all. Gila monsters swallow small bird eggs whole, but they break the large ones and lap up the contents along with mouthfuls of sand. Raccoons, weasels, and many other small mammals bite into bird eggshells and suck out the contents. Some animals scavenge bird eggs that are already broken. Sheathbills (pigeon-sized shorebirds) dart around in penguin colonies searching for abandoned broken eggs. They wedge their bills into the existing cracks and slurp.

Some birds that feed on bird eggs break the rigid eggshells in innovative ways. An Egyptian vulture normally picks up a small egg in its bill and drops it. But the real prize, an ostrich egg, is far too large (about 5.75 by 4.75 inches) and heavy for this technique, and the shell is too strong for the vulture to break with its beak. In some areas of Africa, Egyptian vultures have learned that they can break the thick eggshell by dropping a stone onto the egg. Apparently one vulture "invents" this method and then others copy it. Amazingly, this technique has been invented independently on a different continent by a different bird. Some black-breasted buzzards in Australia smash emu eggs with stones. Emu eggs are just a little smaller than ostrich eggs, at about 5.25 by 3.5 inches. Perhaps there's a yet-to-be-reported behavior of a South American bird that drops stones onto rhea eggs (about the same size as emu eggs).

African egg-eating snakes specialize on bird eggs, and they separate the insides from the shell in a unique way. At night the snakes cruise through the trees searching for nests. Once a snake finds a nest with eggs, it investigates an egg carefully with its tongue and then coils around it to hold it in place. The snake yawns a few times to limber up its jaw muscles, then pushes its mouth onto the egg and gradually wraps its head around the meal. At this point, the snake looks as if it's suffering, with its entire head and throat distended.

These snakes, with bodies the width of your middle finger, can swallow whole eggs three to four times the diameter of their heads. This is roughly equivalent to you swallowing a full-sized watermelon. How can they do it? African egg-eating snakes have unusually elastic skin on the neck, throat, and in the mouth, which gives them a huge gape. Once the egg has traveled a short way down the throat, bony spines projecting into the gullet from the snake's vertebrae puncture the eggshell. The egg collapses. A little farther along, other bony extensions from vertebrae slit the embryo's membranes. The snake swallows the embryo and associated liquids, but because the bony spines prevent the shell from proceeding farther, the snake regurgitates the crushed remains.

THE NEXT TIME you have an omelet or an egg salad sandwich—or caviar, sea urchin eggs, or snail eggs—consider yourself privileged to be in the company of such illustrious egg-eaters as Puerto Rican coquís, bromeliad-living tadpoles, Egyptian vultures, and African egg-eating snakes.

LIFE IN AN ORGANIC SOUP

Gutless Wonder

Though lacking skeletal strengths
Which we associate with most
Large forms, tapeworms go to great lengths
To take the measure of a host.

Monotonous body sections
In a limp mass-production line
Have nervous and excretory connections
And the means to sexually combine

And to coddle countless progeny
But no longer have the guts
To digest for themselves or live free
Or know a meal from soup to nuts.

JOHN M. BURNS, *BioGraffiti*

A tapeworm's life is aptly described in Burns's verse. Disagreeable as these parasitic "gutless wonders" may be, they're prime examples of a lifestyle that permits one animal to live inside another—taking what it needs from the host (to the host's detriment) and giving back nothing positive in return.

Adult tapeworms are ribbon- or tapelike in form; most are white, though some are gray, cream, or yellow. Tapeworms have no skeletal support, no mouth, and no intestine, yet some are more than forty-five feet long. By comparison, *Tyrannosaurus rex* extended a mere forty feet. A tapeworm's body consists of a "head," or scolex, a "neck," and up to three thousand linked blocklike segments called proglottids ("monotonous body sections" in Burns's verse). New proglottids form continually at the base of the neck, like links added to a chain. The newest proglottids are sexually immature, whereas those in the middle of the body are fully functional. The oldest proglottids, farthest from the scolex, are ripe with eggs.

The scolex usually has suckers, a circle of hooks, or both, which attach to the inside of the host's intestine. There the tapeworm remains, anchored to its only source of food. It simply absorbs nutrients through its body wall. Tiny projections cover the surface of each proglottid and increase the absorptive surface area, allowing the tapeworm to take an abundance of nutrients from the rich organic soup.

As Burns notes, these gutless wonders have nervous and excretory connections. The tapeworm's nervous system is composed of a few ganglia (clusters of nerve cells), nerves, and nerve cords. Its excretory system consists of specialized cells and collecting tubules that carry fluids out through the surface of the proglottids. In addition to ridding the body of waste, the tubules help maintain a steady pressure inside the tapeworm, assuring that the animal doesn't deflate or explode.

Tapeworms are basically baby-making machines, facilitated by a complementary set of male and female organs in each proglottid. Often sperm from the same proglottid or an adjacent one ("the boy-and-girl next-door") fertilize the eggs. When two or more tapeworms live in the same area of the intestine, sperm from one animal's proglottids may fertilize the eggs from the other animal's proglottids. Segments full of fertilized eggs separate from each other and either split open and release the eggs inside the intestine or pass out whole in the feces. A single segment may contain up to fifty thousand eggs, and a tapeworm can release half a million eggs each day. Nearly all the eggs will die without ever infecting a new host.

Tapeworms live in at least two different host animals during their life cycles. The egg is swallowed by one animal, where it hatches into a larva. It stays there until a second animal eats the first and inadvertently ingests the larva. Once in its final home, the immature parasite latches on to its new host, eats, matures, and cranks out eggs.

Most tapeworms live only in specific hosts. For example, the sexually mature stage of a beef tapeworm lives only in humans, and the immature stage lives only in cattle. Humans pick up this parasite from eating undercooked hamburger or steak that contains tapeworm larvae. The larvae grow and mature quickly in the person's intestine, reaching a length

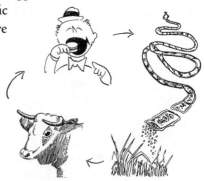

of about twenty feet—nearly as long as the entire human intestine. Eggs pass out from the person in his or her feces. If the eggs end up outdoors instead of in a flush toilet, the embryos can remain viable for up to five months, just waiting for their big chance—grazing cattle. If a steer ingests the eggs, they hatch in the steer's intestine. The larvae penetrate through the intestinal wall, enter the steer's blood vessels, and become lodged in muscle. Steer muscle becomes the beef cuts and hamburger in your grocery store. If you eat the beef without cooking it sufficiently, you help the tapeworms begin their cycle over again. You're not the only one. Beef tapeworms are common parasites, with an estimated 50 million cases of human infection worldwide each year.

Fish-eating birds, mammals (including humans), and carnivorous fish are final hosts for fish tapeworms. Eggs pass out in the feces of the final host's body and hatch if they land in water. Water fleas eat the larvae. Fish eat the infected water fleas, and the ingested tapeworm larvae penetrate through the fish's intestine and settle in its muscle. When a fish-eating animal such as you eats the fish uncooked, the larvae grow into adults in the host's intestine and begin producing eggs, completing the life cycle. (Knowing that, you might wish to reconsider your fondness for sushi.)

Sometimes humans are the intermediate, not final, hosts. Larval tapeworms are much more debilitating than adult tapeworms to humans. With an adult tapeworm, usually you just don't feel well—you may have excessive appetite or loss of appetite, weight loss, diarrhea, or anemia. In contrast, if you're housing tapeworm larvae, they may end up in your eyes, brain, or heart, where they can cause considerable damage.

Consider, for example, pork tapeworms. These tapeworms tend to use humans as the final rather than intermediate host. But not always. Unlike the beef tapeworm eggs, which die if swallowed by a human, when a person eats pork that contains tapeworm eggs, the eggs can survive. After hatching in the intestine, the immature pork tapeworms often enter the person's brain. Within a few years, the person may develop convulsions, paralysis, personality changes, and behaviors that mimic forms of mental illness. If they enter a person's eyes, the immature tapeworms can cause blurred vision or a detached retina. Worldwide, an estimated fifty thousand people die each year from complications related to the immature stage of pork tapeworms. The moral of this story: Thoroughly cook your pork chops and sausage, and don't even *think* about pork sushi!

AFTER READING ABOUT these lovely creatures, you're no doubt hoping to avoid hosting a tapeworm. Isn't our U.S.-inspected beef and pork the safest in the world? Doesn't the inspection process guarantee that the meat we buy is from healthy animals, processed under sanitary conditions, and safe to eat? Why then are we advised against eating undercooked beef and pork?

According to the USDA Meat and Poultry Hotline, laws governing the feeding practices of cattle and pigs have largely eliminated the incidence of tapeworm in the meat we buy in the United States—but not entirely. Inspectors normally wouldn't see tapeworm eggs or larvae unless they cut into the muscle exactly where the eggs or larvae happen to be. Thus, the USDA advises cooking beef to 145°F and pork to 160°F. Tapeworm is a bigger problem in developing countries not because of less stringent inspection procedures, but because of feeding practices. Pork tapeworm is a special problem in areas where pigs are allowed to roam free and eat human feces.

WE THINK OF tapeworms and other parasites as disgusting creatures, but their lifestyle is highly successful. In fact, biologists estimate that nearly half of all animal species live on (ectoparasites) or inside (endoparasites) other animals—a lifestyle that may harm but generally doesn't kill the host.

ANGLING LURES: MASTERS OF DECEPTION

He came so slowly it seemed as if he and history were being made on the way. After a while he got to be ten inches long. He came closer and closer, but beyond a certain point he never got any bigger, so I guess that's how big he was. At what seemed a safe distance, the ten-incher began to circle George's Bobcat Special. I have never seen such large disbelieving eyes in such a little fish. He kept his eyes always on the fly and seemed to let the water circle him around it. Then he turned himself over to gravity and slowly sank. When he got to be about a six-incher he reversed himself and became a ten-incher again to give George's fly a final inspection. Halfway round the circle he took his eye off the fly and saw me and darted out of sight.

NORMAN MACLEAN, *A River Runs through It*

With just a bit of envy and lots of admiration, I used to watch my dad flyfish in the Adirondacks. Like many fishermen, he was passionate about his sport. He had such patience and attention to detail. He wasn't just out for a pleasant afternoon. Dad clearly wanted to outsmart those trout. Fishing for him was an art form that afforded considerable challenge.

The challenge begins with deception—luring a fish to something that looks like, or is, food. The art involves knowing which kinds of natural baits or artificial lures to use in which lakes, rivers, creeks, bays, and oceans, and for which kinds of fish. Worms, salmon eggs, crawfish, crabs, minnows, frogs, salamanders, plugs, spinners, spoons, flies, or jigs. Fish aren't predictable. Although each species has its own personality, individuals can be aggressive or passive. Elusive or cooperative. An accomplished fisherman earns his or her reputation.

Fishermen take pride in their skill and cleverness, but they're not unique in their attempts to deceive prey. Long before humans thought of tricking fish into thinking a bone, plastic, or metal blob attached to a hook was a real meal, some fishes, amphibians, and reptiles were using parts of their own bodies as lures to attract prey. These deceitful predators exploit the responses of their prey to simple cues: when the prey animal sees a small moving object of the appropriate size and shape of a food item, it pounces. Generally this is an effective way for the animal to secure its meal, but occasionally the pounce leads to its own death.

Anglerfish, perhaps the best-known natural anglers, are rather inactive fishes with large heads and cavernous mouths. Their strategy is to camouflage themselves in their environment, where they resemble rocks, sponges, or algal beds and wait for prey to come to them.

Most of the nearly three hundred species of anglerfish have the first ray of the dorsal fin modified into a long, narrow appendage, the "fishing pole," which dangles in front of the fish's large mouth. Some species have spectacular whiplike fishing poles that stretch out to three times the fish's body length. Other "poles" are shorter and less showy. The pink, red, white, or gray lure at the end of the fishing pole is a fleshy structure that the anglerfish moves back and forth evocatively. Depending on the species, the lure resembles an amorphous blob, a worm, a small fish, or a fantasy creature with waving tentacles. When another fish approaches what it assumes to be a meal, the anglerfish suddenly whips away the rod and lure and opens its mouth. Water rushes in and sucks the prey into the orifice. Many anglerfish have sharp teeth that point backward, ensuring that once caught the victim can't escape.

The whiskery batfish (a kind of anglerfish) is covered with outgrowths of skin that resemble bits of seaweed. When a small fish approaches to nibble at the presumed seaweed, the batfish exposes its fishing pole and a lure that mimics a wiggling worm. Now the hungry fish has a choice: seaweed or a worm. The batfish alternately dangles the lure in front of the fish and withdraws it to entice the fish to come within capture range.

Some anglerfish live in deep water, in permanent darkness. There only females have fishing poles and some dangle glow-in-the-dark lures. The light comes from luminescent bacteria living inside the lures. These lures mimic the movements of zooplankton (tiny free-floating animals) and attract small fish and crustaceans to the anglerfish's mouth. Male deep-sea anglerfish not only don't have lures; in many species, they don't even catch their own food.

Population densities of deep-sea anglerfish are low, and mates are hard to locate in perpetual darkness. Some species have evolved an unconventional solution to this problem. Females attract the much smaller males by releasing a pheromone (chemical signal) and by wiggling their luminescent lures. Each species has a distinctive type of lure; thus, a male can recognize a female of his own species. A male in search of a female isn't mature, however, and an advertising female doesn't yet have well-developed gonads. Mere babes in the woods, they are. Once a male locates a female, he bites into her flesh with strong teethlike bones at the tip of his jaws and firmly attaches himself. In time the male's testes mature and the female develops eggs. The dwarf male becomes permanently fused to his giant mate, their circulatory systems unite, and he depends on her for blood transport of nutrients. He has no use for a fishing pole and lure. He's a parasite on his "wife" for life, his body degenerating to a lump of tissue protecting his testes.

Stargazers, named for the eyes atop the head, are one of the ugliest of all fish, with large broad heads and huge mouths. Many species bury themselves in the sand or mud, with only their eyes, nostrils, and part of their mouth exposed, where they wait for food to come to them. They wiggle red or white appendages located under the tongue back and forth in the sand, contracting and expanding what appear to be hyperactive worms. Small crustaceans and fishes are easily fooled into thinking the lures are food.

My favorite luring animal is the South American horned frog, or Pac-Man frog as it's called in the pet trade. Have you played the arcade video game of Pac-Man? In the game, hungry blobs with huge mouths cruise through a maze looking for food. The object of the game is to feed the blobs and keep them from getting eaten by other hungry creatures. Horned frogs are the spitting image of these blobs.

Horns of skin jut out from a horned frog's upper eyelids. Its six- to eight-inch saucer-shaped body lies half buried in mud or leaf litter. There the horned frog waits for unsuspecting prey to amble within range. And then the outrageous mouth, its width more than half the frog's body length, opens and engulfs the lizard, mouse, bird, large insect, or frog—even another horned frog.

Horned frogs also lure other frogs to within striking distance. A horned frog raises a hind foot barely off the ground and alternately vibrates the fourth and fifth toes, which by their very nature resemble worms or insect larvae. Sometimes the frog will go through contortions and raise its hind leg up and over its body so the toes are above its head. In this position, the only way an interested frog can get to its "meal" is to approach the horned frog from the front and face the "wriggling worms" head-on. *Wham!* The horned frog pounces on the duped frog.

Some quite massive animals lure their prey. The largest freshwater turtle in North America, a 150-pound alligator snapping turtle, lies partially buried in organic muck on the bottom of a pond, lake, or slow-

moving river. Its huge shell, overgrown with algae, provides camouflage as it rests motionless, mouth agape. Inside the dark brown mouth lies a forked tongue. The tips of each fork, pink and plump, resemble a pair of juicy worms. The turtle slowly wiggles the tips of its tongue back and forth until a hungry fish can resist no longer and swims into the gaping mouth.

Juvenile Pacific Coast aquatic garter snakes wiggle their tongues to lure young-of-the-year steelhead trout and occasionally young chinook salmon. A snake slowly approaches the edge of a stream, crawls onto a rock, and orients its head less than an inch from the surface of the water. Once settled into its ambush position, the snake waits for a fish to enter into its field of vision. At that point, the snake faces the fish, fully extends its forked tongue, and quivers the tongue tips at the water surface. If the snake is lucky, the fish approaches the "worms" or "insects." Once the fish is within striking distance, the snake makes its move and, if quick enough, catches dinner. Pacific Coast aquatic garter snakes give up fishing as they age. Instead, adults actively forage for Pacific giant salamanders in the streambed.

Some snakes wave or wiggle their tails to attract prey. Luring is most commonly used by pit vipers, such as some rattlesnakes, copperheads, water moccasins, fer-de-lance, and green pit vipers, but also by some sand vipers, death adders, and green tree pythons. The outcome is the same as in fishes, horned frogs, alligator snapping turtles, and aquatic garter snakes: the unsuspecting animal that comes to feed on the presumed worm or caterpillar becomes prey itself.

The luring trick is performed mainly by juvenile snakes because they prey on frogs and lizards, which eat small wiggling prey. As snakes mature, many species switch over to a diet of birds and mammals, which are less attracted to wiggling prey. In luring species, a juvenile's tail is often a conspicuous and contrasting color that then fades as the snake matures.

One of the few snakes that continues to lure when mature is the Saharan sand viper. Both juveniles and adults feed on lizards. A Saharan sand viper's body is the color of sand except for the banded black-and-white tail. A viper spends much of the day buried, with only its eyes and snout peeking above the surface of the sand. If it detects a lizard, however, the snake pokes its banded tail above the sand and slowly wiggles the lure. Snapping at what seems to be a meal, the lizard instead gets struck by the viper, which injects it with paralyzing venom.

Green pit vipers also eat small lizards throughout their lives, not just as juveniles. And both juveniles and adults lure prey with their tails. A green pit viper has yellow stripes running down its light green body, and a pink tail for contrast. When luring, a green pit viper rests on a tree branch, coils its body, sticks the pink tail up through the coil, and twitches it spasmodically. An interested lizard nearby may cock its head, then sprint toward the pink object that resembles an insect larva. The viper strikes the lizard and swallows it whole. Luring allows these arboreal snakes to remain still, camouflaged among the green leaves, and yet feed on highly active prey such as lizards and frogs.

INDEED, A COMMON thread running through all these alluring behaviors is that the angler remains still and blends in with its environment, whether it be water, sand, mud, leaf litter, or vegetation. So, too, a human angler knows that camouflage, silence, and patience are all crucial to a successful catch. If the angler is lucky, the interested party will circle one more time around even George's Bobcat Special, throw caution to the wind, and bite.

STOMPING FOR WORMS

All around, worms are poking their angry pink heads up through the blackened soil. They shoulder themselves out of their burrows, shimmying, dancing, twitching along on their ten-inch bellies, driven to frenzy by this infernal vibration. Then Hill, a forty-year-old who learned worm grunting at his daddy's knee, grabs a paint can and strides around snatching up his haul, about 160 worms in five minutes. They will do their final dance on a fishhook. . . .

RICHARD CONNIFF, *Spineless Wonders*

Ruben Hill has been hammering on a stake driven into the ground—
"grunting" for earthworms in the Florida Panhandle. Vibrations from the hammering cause the worms to surface, allowing Ruben to collect five thousand worms in half a day. He'll net $150 when he sells them to local bait shops. Not bad for half a day's work. Ruben can take the rest of the day off and go fishing.

That's one way to grunt for worms. Other simple methods involve driving a wooden stake into the ground and drawing a notched piece of wood back and forth across it like a saw, or sticking a pitchfork into the ground and hitting a prong with a metal object. Professionals engaged in the multimillion-dollar worm-grunting industry in the United States often use electric probes. A concrete contractor devised what he calls the "Mother of All Worm Grunters." He drove a five-foot length of galvanized pipe into the ground, leaving about eight inches exposed. Whenever he needs worms for fishing bait or as a treat for his chickens, he inserts his concrete vibrator into the pipe and lets it run about thirty minutes. The worms surface en masse, shimmying, dancing, and twitching.

Every year since 1980, Nantwich, in Cheshire, England, has hosted the World Worm Charming Championships. ("Charming" is the English's genteel term for what North Americans call "grunting.") The person who charms the most worms from his or her three-meter square plot of ground in thirty minutes earns the golden worm-shaped trophy. The participant charming the heaviest worm wins the silver worm-shaped trophy. Titles are held for one year, after which the trophies are relinquished to the new champions, who likely trained year-round for the event. Strict rules, translated into thirty-one languages, regulate how earthworms may be charmed.

- No drugs to be used! Water is considered to be a drug/stimulant.
- Any form of music may be used to charm the worms out of the earth.
- A garden fork of normal type may be stuck into the ground and vibrated by any manual means to encourage worms to the surface. . . .
- A piece of wood, smooth or notched[,] may be used to strike or "fiddle" the handle of the garden fork to assist vibration.

- Competitors who do not wish to handle worms may appoint a second to do so. The second shall be known as a "Gillie." . . .
- Charmed worms to be released after the birds have gone to roost on the evening of the event.

(The complete set of eighteen rules is available on http://mysite.freeserve .com/wormcharming/ should you wish to participate.)

WHY DO VIBRATIONS drive earthworms to the surface? We don't know for sure. Possibly vibrations caused by any object rhythmically affecting the ground mimic those caused by raindrops. Earthworms come aboveground during a hard rain to avoid drowning, so their reaction might be a generalized response to avoid suffocation.

Alternatively, earthworms might move up and out in response to vibrations as an innate escape tactic from their arch enemies: moles, which have enormous appetites packed into their small bodies. Moles can eat their weight in earthworms every day; these mammals spend much of their lives tunneling through the soil in search of earthworms to fuel their high metabolic rates.

Moles often hoard extra earthworms once they're satisfied for the moment. From late fall to early spring, European moles store hundreds of worms in caches in their nest chambers or embedded in the tunnel walls of galleries leading out from their nests. A mole eats or mutilates the first few segments of an earthworm, rendering it immobile. Mutilated earthworms stay fresh and otherwise healthy for months when the underground temperatures are cool. There's one problem, however. Earthworms gradually regenerate their damaged bodies and wriggle away once the soil warms in the spring.

SOME NONHUMAN ANIMALS also urge earthworms to the surface by thumping the ground. I remember vividly the day twenty years ago when a friend and colleague at the University of Florida, Jack Kaufmann, told me about the amazing discovery he made while studying wood turtles in the mountains of central Pennsylvania. Wood turtles have rough-textured shells about five to eight inches long, orange blotches on their brown necks and legs, and long tails. At home in the water, they spend the winter hibernating in creek bottoms. During warm months, they hang out around the creeks and eat tadpoles, snails, and small fish. They also wander through woods and farmlands, where they feed on mushrooms, green leaves, fruit, and insects. And earthworms.

Jack discovered that wood turtles catch earthworms by stomping their front feet on the ground. He could hardly believe his eyes the first time he watched this behavior. A turtle typically stomped with one front foot several times, at a rate of about one stomp per second. It then stomped several times with the other front foot. It either repeated alternate foot stomping or moved elsewhere and tried again. Whenever an earthworm appeared, the turtle seized and swallowed it. To convince himself that the turtles' stomping actually caused the worms to surface, Jack imitated the behavior by tapping the ground with two fingers with the same force and rhythm the turtles used. Sure enough, worms appeared. Jack observed worm stomping sixty-three times during 830 hours of watching the turtles. Most stomping sessions lasted at least fifteen minutes, and one persistent turtle continued for over four hours.

How successful is the wood turtles' technique? During sessions where Jack watched turtles stomp for at least one hour, they averaged 2.4 worms per hour. One turtle caught fifteen worms within an hour. One fellow (was he a pro or just lucky?) caught seven worms within eight minutes. And Jack? During successful tapping sessions, he averaged one worm every nine or ten minutes—better than some turtles, though not as skilled as others.

Jack encountered skepticism when he submitted his manuscript for publication. Not all the manuscript reviewers believed that turtles could exhibit such behavior. One reviewer suggested that the observations of foot stomping be published, but without any mention of the earthworms! Nevertheless, natural history won out. The paper was published in a prominent herpetological journal, and other field biologists have since corroborated the behavior.

Some birds also capitalize on earthworms' response to vibrations. Herring gulls and some other gulls "paddle" for earthworms. A gull lifts its legs alternately and beats its feet against the ground at a rate of about four beats per second. Moving its head rhythmically back and forth, the bird watches for earthworms that wriggle onto the surface. Lapwings and other plovers have their own unique way of encouraging earthworms to surface. A bird stands on one leg and then rapidly quivers its other outstretched leg. The earth beneath trembles and earthworms seemingly

appear from nowhere. Have you ever watched
American robins hunt for earthworms? They
sometimes thump the ground with their feet
before cocking their heads and pulling and
stretching their reluctant prey from the ground.

Perhaps animals other than humans, turtles, and birds use vibrations
to catch earthworms, and we just don't know about them yet. We should
look for such behavior in worm-eating mammals such as pigs, foxes, and
badgers, which feed aboveground. Who knows—perhaps they stomp their
feet, thump their snouts, or bang their tails against the soil to coax earthworms to the surface.

THE NEXT TIME you need some fishing bait, remember Ruben Hill's
strategy. Get yourself a stake and a hammer . . . and then pound. It's a skill
worth developing. And if you get really good at this, consider a try for the
World Worm Charming Championship in England.

A TEAM EFFORT: HOW (SOME) ANTS GET FOOD

Imagine sawing off a barn-door at the top of a giant sequoia, growing at the bottom of the Grand Cañon, and then, with five or six children clinging to it, de-
scending the tree, and carrying it up the cañon walls against a subway rush of
rude people, who elbowed and pushed blindly against you. This is what hundreds of leaf-cutting ants accomplish daily, when cutting leaves from a tall bush,
at the foot of the bank near the laboratory.

WILLIAM BEEBE, *Edge of the Jungle*

Ants, the most highly developed of all social insects, live in organized
communities, or colonies. Most ant colonies consist of three castes:
queens, males, and workers. An ant colony includes at least one queen,

whose main occupation is laying eggs. She's gen-
erally the largest ant in the colony. Most of the
other ants are workers, all of whom are females
incapable of reproducing. In many species of
ants, workers vary in size and carry out different
chores. Large soldiers may defend the colony;
smaller workers gather food and feed the queen.
Males are usually intermediate in size between the queen and the workers.
They hatch and reach adulthood only at certain times and do no work at
all; their only occupation is to fly away at the right moment and mate with

young queens just emerging from other colonies. Then they die . . . a short life and a merry one.

Let's look at one aspect of colonial life: cooperation in securing food. Ants as a group eat a wide range of food, and they exhibit extraordinary behaviors to get what they eat. The following are just a few.

AS WILLIAM BEEBE observes, New World leaf-cutting ants are remarkably strong. Workers have a complex division of labor depending on their size. Intermediate-sized workers called medias snip pieces of leaves many times their own size. Nicknamed parasol ants, they grip the leaf fragments between their sawtoothed mandibles and carry their burdens high above their heads as they march back to their huge underground nest. Why cut and carry leaves? Not to eat, not to roof their nest, and not to provide warmth once the leaves decay—all of which were once considered possibilities. Instead, the leaves provide the compost for growing what the ants do eat: a specialized fungus. The ants eat the small terminal bulbs on the hyphae (fungal threads) that grow on the leaf substrate. They also feed the bulbs to their larvae and to the queen.

Back at the nest, smaller media workers tend the fungus gardens. They chew the newly transported leaves, inoculate the saliva-laden mush with the fungus, and fertilize the leaf mush with feces. Another chore is dumping dead fungus, dead ants, and other refuse onto a rubbish heap outside the nest.

These smaller media workers also protect their crop from attack by a very different type of fungus—a parasitic mold that is so deadly it can wipe out a garden of the edible fungus in a few days. The ants' weapon is a mold-killing bacterium. The ants carry colonies of the bacterium either just below their mouths, on what would be their necks if they had necks, or behind their front legs. Whenever they encounter the deadly mold, the ants rub their mold-killing bacterium on the spot. Antibiotics produced by the bacterium inhibit growth of the mold. Cameron Currie, from the University of Kansas, and his collaborators believe that this amazing four-part interaction of ant, edible fungus, parasitic mold, and bacterium has existed for at least the past 50 million years—for as long as leaf-cutting ants have been gardening.

Leaf-cutting ant workers come in two other sizes. Maximas, the largest, have extra-large jaws; they protect the nest from intruders. These soldiers usually stay inside the nest entrance, leaving only to defend the colony. Minimas, the smallest workers, tend the baby ants inside the nest; they also weed the garden and harvest the fungus bulbs. Some of these pint-sized ants ride outdoors on the leaf fragments carried by the medias, where they protect their preoccupied nest mates from parasitic flies that try to lay eggs on them. The minimas wave their legs about and snap their mandibles at the flies.

The edible fungus is a precious commodity. When a young queen leaf-cutting ant leaves the nest to form a new colony, she carries a wad of hyphae from her colony's fungus garden in a small pocket in her mouth. After mating, she digs the beginning of a nest and then fertilizes the hyphae with her feces—the beginning of a new fungus garden. Nearby she lays her first eggs. Until these young pupate and emerge as workers forty to sixty days later, the queen does all the gardening herself.

Several million leaf-cutting ants may live in one colony. A nest may be over three hundred feet wide and reach twenty feet down into the earth. One nest excavated in Brazil housed more than a thousand chambers, of which 390 contained fungus gardens. Worker ants had moved about forty-four tons of earth to dig nest chambers. Ant experts equate the building of a leaf-cutter nest of this size, in human terms, to constructing the 4,000-mile Great Wall of China, built of stone and brick entirely by hand.

ARMY ANTS, FROM the New and Old World tropics, are sometimes described as "the Huns and Tartars of the insect world." As many as a half million worker ants hunt together for food by marching in columns or swarms. They bite or sting just about any encountered creature that can't escape: tarantulas, scorpions, beetles, ants, cockroaches, grasshoppers, and even small snakes and lizards and nestling birds. Rather than eating on the spot, the ants sling their booty under their bodies and between their legs and carry them back to the nest, where they'll be devoured by themselves and their nest mates, stored, or fed to the larvae. When a captured prey is too bulky or heavy for a single ant to transport, a team of workers drags or carries the victim or cuts the animal into small pieces. Two or more workers often run in tandem, carrying elongated prey such as centipedes and beetle grubs. Large soldier ants, with huge and powerful sickle-shaped mandibles, march on both sides of the swarm and protect their smaller worker nest mates as they capture and transport prey.

The best-studied army ant is a swarm raider, *Eciton burchelli,* a common species from Mexico to Paraguay. These ants don't build nests; instead, they live in temporary camps in sheltered locations. They bivouac each night by creating shelters out of their own bodies linked together by hooked claws at the tips of their six feet. Up to a half million workers' bodies protect the queen and her thousands of youngsters, creating a musky-smelling mass. At daybreak, the workers disentangle their bodies, and with powerful jaws and deadly stingers poised for action, they begin their day's raid. The swarm strikes out in a different direction each morning, as the previous day's area is now largely depleted of food.

Eciton burchelli ants alternate between two phases. During its static period, the colony bivouacs at the same site for two to three weeks. Soon after arriving at a new site, the queen develops and lays about 60,000 eggs. About midway in the static period, she lays another 100,000 to 300,000 eggs. Late in this phase, larvae hatch from the first batch of eggs; the larvae form cocoons, and then several days later they emerge from their cocoons en masse—ready to work. The growing size of the raiding swarm and the increased level of activity pushes the colony into its nomadic phase, during which the ants migrate to a new bivouac site at the end of each day's raid. This phase usually lasts from two to three weeks, as long as the larvae from that second batch of eggs are growing and eating. Once these larvae spin cocoons, the colony stops migrating and settles back into its static phase.

The interests of army ants and humans sometimes collide. My first day in the Brazilian rain forest, I accidentally stood in the path of a raiding swarm of *Eciton burchelli;* dozens of irate ants crawled up my boots and inside my jeans, where they took revenge on my bare skin. I did exactly what you would have done. After dancing around hysterically for a minute, I stripped and frantically brushed the ants off my bare body.

On the positive side, humans in some areas of South America exploit army ants' powerful jaws: they use them as butterfly stitches to suture open wounds. One person holds the edges of the wound together, then a second person allows angry ants to bite along the length of the wound. Once an ant has sunk in its mandibles, the "surgeons" cut its body off, leaving only the head and jaws. The patient leaves the "sutures" in place until the wound heals—an efficient, cost-effective treatment.

Carl Stephenson immortalized army ants in his 1938 short story "Leiningen versus the Ants," in which a Brazilian plantation owner fights an advancing swarm of ravenous army ants. Hollywood filmed this tale of man against nature: *The Naked Jungle,* starring Charlton Heston and Eleanor Parker (1954). The year is 1901 and the *marabunta* (plague of ants), a swarm of agonizing death twenty miles long and two miles wide, is advancing through the jungle, swallowing everything in its path. Heston, the cacao plantation owner, refuses to flee. He builds a moat around his mansion, but when the ants cross the water on leaf fragments, he tosses furniture out and torches it to keep the ants at bay. When fire doesn't repel the ants, Heston blows up a dam and floods his paradise. As the ants drown in the rush of water, Heston declares he is giving back everything he took from the river. The footage of the boiling *marabunta* is frightening, as is the image of ants swarming over human bodies, stripping the flesh to skeletons. It's an entertaining movie, even though Hollywood filmed relatively innocent, vegetarian carpenter ants instead of *Eciton burchelli*.

HARVESTING ANTS GATHER seeds from deserts and semideserts around the world. The worker ants carry their cargo back to their nests, where they bite off the radicles (the future roots of the embryos) of their harvested seeds to prevent them from germinating. Then they store the seeds in underground granaries—up to four hundred chambers in a single nest. The ants gather seeds during the warm part of the year, ensuring they'll have enough food for winter. In some species, if their seeds get wet, the worker ants haul them out into the sun to dry.

Different individuals perform different chores. The largest workers usually defend the nest and food cache. In some species, massive-headed workers take on the job of milling the seeds: stripping away the outer coat and breaking apart the endosperm (nutritive tissue surrounding the embryo) with their large blunt-edged mandibles. Smaller workers do housekeeping tasks; care for the queen, eggs, and larvae; scout for seeds; and gather seeds. The system wouldn't function efficiently without every worker doing her job.

The industriousness of harvester ants has long been appreciated. In one of his fables, Aesop told about the lazy grasshopper who sang and frolicked all summer, stored no food for winter, and then begged the industrious ants for some of their store of grain. The moral: Work hard and prepare for the future. Solomon likewise praised harvesting ants (Proverbs 6:6–8) and encouraged people to learn from them: "Go to the ant, O sluggard; consider her ways, and be wise. Without having any chief, officer or ruler, she prepares her food in summer, and gathers her sustenance in harvest."

Harvesting ants exert power over the plants near their nests, influencing which ones flourish or flounder. They prefer to eat some seeds over others, so they harvest some and leave others to germinate. But even preferred seeds may sometimes gain from the ants' activity. Seeds may be dispersed farther away from the parent plant than would otherwise be possible because careless workers drop seeds before returning home, and ants in a hurry may neglect to chew off the radicles, allowing the seeds to germinate inside the nest.

A MUTUALLY BENEFICIAL interaction, found both in the New and Old Worlds, occurs between ants and certain sap-sucking insects such as aphids (plant lice). These ants feed on a sweet liquid called honeydew excreted by the sap-suckers. Just as we tend domestic cattle for milk, some ants tend herds of aphids from which they solicit honeydew, earning themselves the nickname "dairying ant." A worker dairying ant "milks" an aphid by stroking it with her antennae or front legs, and the aphid responds by releasing a droplet of honeydew from its rear end. The ant then goes to the next aphid and solicits more sugary liquid. She milks as many aphids as necessary to fill her abdomen, then returns to her nest, where she regurgitates for her nest mates who have stayed behind to perform the day's housekeeping and care-giving chores.

What do the aphids get in return? Protection from aphid-eating enemies such as beetles and lacewing larvae. The ants also drive off parasitic wasps and flies before they can lay their eggs on the aphids. Some kinds of dairying ants carry their charges to new less-crowded foliage or to places where the ants can provide better protection. One species of North American dairying ant keeps the eggs of the American corn-root aphid in its nest over the winter. Once temperatures warm in the spring, the ants put the newly hatched young out to pasture: they carry them to the roots of nearby food plants. If any of those plants die, the caretakers move the aphids to healthy plants.

WORKER HONEYPOT ANTS from the deserts of the New and Old Worlds gather termites, plant nectar, and honeydew from insects. They return to their nest and feed these delicacies to nest mates, called repletes, usually the largest workers. The repletes swell up to the size of large peas, and their abdomens are so bloated with a cocktail of liquefied termite protein, nectar, honeydew, and water that they can barely walk. They climb to the chamber ceiling and hang by their claws, where they live as airtight tanks of stored food. During droughts, when food is scarce, nest mates request meals by tapping them with their antennae, and the repletes regurgitate their sweet

ambrosia. A colony of fifteen thousand ants maintains about two thousand repletes.

Honeypot ants aren't the only ones that benefit from these storage tanks of sweet syrup. In Mexico and in the southwest United States, people nip off the syrup-filled abdomens of repletes as a sweet treat. Australian Aborigines likewise consider repletes a delicacy.

Honeypot ants are pugnacious. Once a larger colony has determined its dominance, it may wage all-out battle on a neighboring weaker colony. The ants force their way into the foreign nest, where they kill the queen and capture her larvae and pupae for slaves in their own nest. In their new nest, the slaves act as though they were sisters of their captors; they perform tasks as they have been programmed. The raiders also dislodge and drag away engorged repletes—extra food during lean times.

ANTS AROSE ABOUT 100 million years ago, during the Cretaceous era, and they shared the earth with dinosaurs. Dinosaurs went extinct, but the approximately ninety-five hundred recognized species of ants (and the thousands yet to be named) have carved out unique niches for themselves and as a group have become abundant in most terrestrial landscapes. Ants' success is due to their strength in numbers; each ant works for the colony, the superorganism.

But don't think of ants as individual organisms altruistically working for the good of each other. They're not. Think of them as detached hands, crawling about doing tasks for the stay-at-home body. Ants are programmed to do their chores based on their physical characteristics, size, and only the rare males and queens are individuals in the sense that they can reproduce.

The 1998 animated movie *Antz*, though an enjoyable tale, incorrectly portrays ants as individual organisms. A little worker ant named Z is a soil relocation engineer, but moving around dirt all day is not his idea of a rewarding career (remember that in real life, all workers are actually female). Not wanting to live, work, and die for the colony, Z is an independent ant. He falls in love with the Royal Princess Bala, and together they leave the colony and seek Insectopia, where the streets are paved with food and no one tells an ant what to do. After various adventures, they return to the colony and save their nest mates from General Mandible (the corrupt head of the soldier caste) and his plan of killing the queen and all the workers, marrying Princess Bala, and establishing a new colony. Z decides that colony life is okay after all, now that it's his choice.

Of course, ant life isn't like that. Ants, which are almost all females, don't really make choices in how and whether to cooperate in colony life. From my way of thinking, *Antz* gives a picture less intriguing than reality. With their programmed division of labor, ants can grow fungus gardens, kill and transport large prey, tap their nest mates for food, and gather food in other diverse ways that would be impossible for solitary individuals.

4 Don't Tread on Me

IT'S A BASIC fact of life: predators seek prey for food, and prey try to avoid being eaten. The best way to avoid being eaten is simply to escape being noticed and identified as edible. If noticed by the predator, however, the animal might still be able to evade capture or defend itself. In this chapter, we'll look at ways prey avoid being eaten through each of these three strategies: escaping notice, evading, and defending. Again, I've chosen unusual examples and have described only a few of the many intriguing ways animals protect themselves.

Some traits of prey lessen the chance of a predator encountering the animal, or at least recognizing it as food. Camouflage (crypsis) enables some animals to blend into their surroundings. Other tasty prey mimic some inedible part of the environment, such as a twig or rock. But predators aren't always fooled. They're adept at spotting deception. After all, if a predator can't find and capture prey, it won't survive.

If a predator recognizes a prey for what it is, escape is the next line of action. The prey animal may attempt to startle or confuse the predator, or try to flee by swimming, flying, running, jumping, or slithering. It may warn the predator: "Don't tread on me!" Skunks stomp their feet, lift their tails, and hiss to warn that they're loaded and ready to fire. Rattlesnakes vibrate their rattles to warn of their lethal venom.

When all else has failed and the prey is in the clutches of a predator, it may try to defend itself—either actively or passively. Last-ditch efforts of many reptiles, birds, and mammals include biting or scratching— or using their weapons, such as the skunk's smelly spray or the rattlesnake's venom. Toads and horned lizards fill their lungs with air, becoming larger and harder to handle. Opossums and other animals feign death "in hopes" that attackers prefer not to eat prey already dead. Hognose snakes roll onto their backs, open their mouths, convulse for a few seconds, and then freeze. Toadfish flip into death throes: they honk, appear to suffocate, and lose color.

We'll begin by looking at two specific animals, each with diverse ways of avoiding being eaten: octopuses and horned lizards. Next we'll look at some unusual anti-predator defenses. Some animals self-amputate body parts when attacked. Others use chemical warfare by spitting or spraying foul substances or by injecting venom through fangs, spines, stingers, or stinging cells. Some animals borrow or steal other animals' defenses, and some decorate their bodies with foreign material.

CHAMELEONS OF THE SEA

Before my gaze the rock started to melt, began to ooze at the sides like a candle that had become too hot. . . . It was my first acquaintance with a live, full-grown octopus. The beast flowed down the remainder of the boulder, so closely did its flesh adhere to the stone, and then slowly, with tentacles spread slightly apart, slithered into a crevasse nearby. . . . It seemed somewhat irritated at my disturbing it, for it rapidly flushed from pebbled yellow to mottled brown and then back to a livid white. It remained white for about twenty seconds and then altered slowly to a dark grey edged with maroon.

GILBERT C. KLINGEL, *Inagua*

Stories and movies often portray octopuses as evil creatures attempting to drag ships down to the sea's bottom or as belligerent contestants in fierce battles against superhumans, who inevitably cut off the octopuses' arms and strew them across the ocean floor. Despite what Victor Hugo, Jules Verne, and Hollywood would have us believe, octopuses are generally shy, gentle animals that tend to hide rather than pick a fight. Occasionally divers get enveloped by octopuses, whose sucker-clad eight muscular arms explore every curve of their bodies. But these octopuses aren't aggressive. They're simply investigating an unknown object. If the person stays calm and still, the octopus will feel around for a while, decide the human isn't edible, and release its grip.

As Victor Hugo wrote in *Toilers of the Sea*, "To believe in the octopus, one must have seen it." The essence of an octopus is a head with well-developed eyes and a large brain, a soft globular bag called the mantle packed with internal organs, a funnel-shaped siphon for expelling water, and eight long arms. Webs of skin unite the arms at their bases. The mouth sits beneath the head at the base of the eight arms. Octopuses seem alien to us, and as such they're shrouded in mystery. They're wondrous enigmas: flabby, boneless, and shell-less distant relatives of clams and oysters (all belong to the phylum Mollusca), designed to survive the rigors of ocean life. These playful, intelligent animals appear almost humanlike as their large eyes make contact with yours and follow your every move. Although they appear alien, octopuses are from this world—just part of it we don't understand very well, the sea. The approximately one hundred species range in size, from a tiny octopus in the Indian Ocean measuring one inch from arm tip to arm tip and weighing less than an ounce, to the giant octopus of the Pacific Ocean, which can grow to more than twenty feet and weigh up to 110 pounds.

An octopus skillfully moves about by strong rhythmic muscular contractions and by jet propulsion. It propels itself by sucking water into its body, then expelling water through the siphon, which causes the animal to shoot through its surroundings—usually backward. Several hundred round, muscular sucking disks on each arm's underside provide strong suction that allows the octopus to attach firmly to rocks. The arms move separately in any direction or work together to crawl along the ocean floor. This high level of coordination needed to manage eight independent arms requires a highly developed brain, making the octopus the most intelligent invertebrate known— at least as smart as the average house cat.

Octopuses are predators, and they sometimes ambush or actively pursue prey. They also pounce on rocks, spread their arms and webs of skin, and search for food with their arm tips. If their arms and webs can't cover the rocks, they probe their sucker-clad arms into blind crevices. The suckers, softer and more flexible than our fingers, easily distinguish textures and are more sensitive than our taste buds to certain tastes. Once the suckers have captured or immobilized a clam or crab and brought it to the mouth, the parrotlike beak cracks it open. Next the octopus injects its prey with neurotoxin from its modified salivary glands. Octopuses prefer to eat in privacy, so they often carry meals back to their dens, usually small caves behind rocks or in reefs. The webs of skin provide baskets, allowing a large octopus to carry a dozen or more paralyzed crabs at a time. The entrance to an octopus's den is usually strewn with a garbage pile of shells and other remains of past meals.

Octopuses have their own predators. Moray eels, sharks, sea lions, seals, otters, elephant seals, whales, and dolphins find octopus meat espe-

cially tasty. To protect themselves from becoming someone else's meal, octopuses employ a complex arsenal of defenses. When detected by a potential predator, an octopus can quickly swim away or launch itself backward.
Often it jets away from the predator and dives into the sand, burying itself. It may discharge a cloud of blackish or dark purplish fluid from its ink
sac and then hide behind the smoke screen. Sometimes the ink cloud itself
resembles the blobby shape of an octopus and acts as a decoy, allowing the
octopus to escape while the predator eyes the blob. An octopus may spread
its arms and webs, appearing larger, and it may threaten by lashing out its
arms and blowing water. Some species display ocelli (false eyespots), giving the impression of a huge animal with frighteningly large eyes.

The surest way of not getting eaten is to avoid detection, to become invisible. Octopuses have been described as "chameleons of the sea," but
they make chameleon lizards look amateurish. To be camouflaged, most
animals must settle on a single appropriate background color or pattern.
A brown toad squats on dirt. A green treefrog perches on a green leaf. Octopuses do the opposite. They settle on or cruise through any background
and rapidly change color and pattern to match their surroundings. Since
the focus of this chapter is defense, we'll look at camouflage as an anti-
predator mechanism. But the octopus's ability to blend into its environment also allows it to sneak up on its own prey and increases its effective-
ness as a predator.

Octopuses do more than just blend in with their environment; they
rapidly change appearance. In describing a favorite octopus in his book
Inagua, Gilbert Klingel writes: "No schoolgirl with her first love was ever
subjected to a more rapid or recurring course of excited flushes than this
particular octopus." This octopus's fashion statement included wide bands
of maroon and cream, wavy lines of lavender and deep rose, red spots, and
irregular purple polka dots. When attack is imminent, many octopuses
suddenly blanch white or darken, whichever is opposite from their previous state. Color change may so startle or confuse the predator that the
octopus can escape before the predator knows what's happening.

Scientists have assumed that because octopuses have soft, tasty bodies,
they probably remain cryptic most of the time they're active and away from
their dens. Roger Hanlon, from the Marine Biological Laboratory, Woods
Hole, Massachusetts, and his colleagues recently studied a diurnal octopus (*Octopus cyanea*) on coral reefs in Tahiti and Palau in the Indo-Pacific.
They found that these octopuses spend less time being cryptic than expected. The octopuses in Tahiti were highly cryptic 54 percent of the time,
and those on Palau only 31 percent.

The videotaped octopuses changed colors and patterns an average of 2.95 times per minute, or 177 times per hour, as they slowly foraged for food. Some of these changes allowed the animals to blend in with their surroundings. One remarkable animal changed its appearance eleven times in sixteen seconds. The octopuses often resembled inanimate objects. When perched on a promontory, octopuses often looked like rounded coral or rocks as they hunkered with their heads down. They also resembled clumps of algae or highly textured rocks by assuming the same color and texture.

When crossing open spaces, the filmed octopuses sometimes engaged in "moving rock" behavior. They assumed the pattern, shape, texture, and color of rocks in their habitat; then with heads flattened and webs spread, they slowly moved across the substrate on the tips of their arms. The animals gauged speed of wave action and other movements in the environment—a behavior that no doubt allows their tiptoed crawls to go unnoticed.

But the octopuses often looked conspicuous, especially when they moved quickly. Since a predator keys in on movement, it might pay an octopus to look like something other than an octopus. Two conspicuous but un-octopus-like patterns included waves of color flowing over the body and dazzling white longitudinal stripes running from the dark mantle onto the dark arms. Octopuses sometimes conspicuously moved across open expanses by swimming forward, with their arms trailing behind. To the human observers, they resembled parrot fish.

These coral reef octopuses combined crypsis, conspicuousness, and mimicry. Why didn't they just stay cryptic? Hanlon and his colleagues suggested that it may be too "neurophysiologically expensive" to change appearance continually in their diverse habitat. Another possibility is that the variety of patterns may prevent a predator from developing a search image—a set of key characters identifying the animal as a particular prey item. Essentially, the octopus may be producing false search images. Think of it this way. If you dash into your local convenience store for a Coke, you look for a red-and-silver can. That's your search image. If Coke came in a variety of sizes, shapes, and colors of cans, it would be much harder to identify the can on the shelf. And worse still if the cans changed color and design in front of your eyes!

Octopuses change color and pattern through their hundreds of thousands of chromatophores, cells that contain single-colored pigments: yellow, orange, red, brown, and black. Muscles surround each chromatophore. A chromatophore expands when the muscles contract, producing color. A chromatophore retracts and returns to its original size when the muscles relax; the effect is that the patches of color disappear. Because the muscles are under nervous control, direct from the brain, the chromatophores expand and retract rapidly. Some chromatophores can be expanded while others stay retracted, resulting in an infinite number of possible patterns. The muscles also enable octopuses to change their skin texture from smooth to bumpy.

An octopus's defenses may be effective against some of its predators, such as marine mammals and sharks, but they don't work well against humans, who have long considered octopus meat a delicacy. Ancient Greeks and Romans considered octopus to be the finest seafood one could eat. Japanese barbecue octopus tentacles on skewers. Mediterranean cooks simmer octopus in a stew with tomatoes, onion, garlic, white wine, and spices. Italians cook octopus in a sauce of olive oil, tomatoes, parsley, garlic, and spices and serve it over pasta. And Latin Americans marinate octopus tentacles in lemon juice, onions, and spices in a *cebiche* cocktail.

Unfortunately, octopuses have no defense against human hunters. Just as curiosity kills the cat, so it kills the octopus. Naturally inquisitive, octopuses explore secluded crannies to use as dens. Four thousand years ago, Egyptians fished for octopus in the same way fishermen do today—by submerging jars, pots, and kegs into the ocean and waiting for octopuses to climb inside. A string of empty beer bottles catches tiny octopuses. So possessive is the octopus of its discovery that it clings to the inside as the keg or beer bottle is drawn back up. Fortunately, octopus mariculture off the coast of Japan, in the Mediterranean, and off the northwest coast of Africa takes some of the hunting pressure off wild octopus populations.

EVEN MORE THAN their taste or fluid movement, perhaps it's the graceful shape of the octopus that has endeared it to people. Ancient Mediterranean artists arranged octopus figures on vases with obvious admiration. The classical Greeks likewise embellished their vessels, shields, and coins with elegant octopus forms. Exquisite Roman mosaics depict octopuses in exact detail. Today, octopus biologists consider their study animals to be among the most beautiful of all living creatures. Until recently, however, we had no idea that octopuses were such talented chameleons and shape-shifters—just another way they defend themselves against their predators.

TEARS OF BLOOD

One remarkable feat of Texas, coast, and regal horned lizards is their ability, when provoked, to shoot narrow streams of blood from their eyes. They shoot blood so infrequently at humans, and reports of it sound so unbelievable, that many people are misled into regarding blood squirting as a fable. But in Mexico people regard them as sacred toads, because, according to folklore, when they cry they weep tears of blood.

WADE C. SHERBROOKE, *Horned Lizards*

Marilyn, our realtor, drove us north past the Flagstaff city limits, through seven miles of ponderosa pines, tall grasses, and golden-yellow sunflowers, and then turned abruptly west and headed toward the San Francisco Peaks, the highest mountains in Arizona. Entering a wide driveway, she announced, "Here's the Bean Farm, site of the original homestead back in the late 1890s."

The old stone barn—built in the 1930s for storing potatoes and pinto beans and now housing a pool table, garage, guest room, and a large room perfect for a study—was tempting enough. But what really sold us on the property was the little brown, spiny horned lizard that looked back at us with beady black eyes as we opened the car doors and stepped out. "Hey!" I blurted. "We'd have our own population of horned lizards here!" My husband was as thrilled as I. Marilyn must have thought our response a bit peculiar, but, eager to make a sale, she tried her best to match our enthusiasm for the lizard. Two days later, believing the horned lizard to be a good omen, we bought Bean Farm and acquired the local population of horned lizards.

I later learned that other people consider these lizards to be a good omen. People along the Texas-Mexican border value and respect horned lizards because they believe the lizards bring much-needed rain. On the Navajo Reservation in Arizona and New Mexico, children hold horned lizards to their hearts and murmur, *"Yáat'ééh shi che,"* which means, "Hello, my grandfather." This greeting shows the lizards respect, and in return the lizards confer strength to the children. An apparent variation of this is expressed by children living in and around Flagstaff who press horned lizards to their hearts and make wishes in the belief that—if they're good—the lizards will grant their wishes. Hundreds of years before

Christopher Columbus "discovered" the New World, Native Americans in what would be the southwestern United States were painting horned lizards on their pottery, carving horned lizard fetishes from stone, creating horned lizard petroglyphs around their homes, and singing and telling stories about horned lizards. Their descendants carry on their belief that these lizards have the power to influence health and happiness.

So why was I, raised in upstate New York and Pennsylvania, so excited about horned lizards? Why do many people find them charming? They're unusual. Short legs and spiny tails extend out from their broad, flat bodies, rather like pancakes fringed with sharp spines. Their heads sprout wicked-looking horns. Horned lizards' squat shape superficially resembles toads, which explains why some people call them horned toads or horny toads. In fact, the genus name, *Phrynosoma,* means "toad-bodied." I think they more closely resemble scaled-down armored dinosaurs.

A SPINY PANCAKE with short legs has a fairly awkward gait, but horned lizards have numerous defenses against their enemies, which include rattlesnakes, collared lizards, leopard lizards, roadrunners, hawks, owls, house cats, coyotes, dogs, and grasshopper mice. The defenses allow a lizard to avoid being detected, to escape if detected, and to resist attack if it can't escape.

First, horned lizards freeze in their tracks and rely on camouflage— they're masters at blending in with their surroundings. Short-horned lizards, the species common around my home, vary in color and often match the Flagstaff volcanic cinders, worn-out farm soil, or the mottled ground under ponderosa forest, piñon-juniper scrub, or aspen groves. A frightened horned lizard flattens its body and waits for danger to pass. This works well against my husband, who rarely ever sees them, even though he's a field biologist.

If camouflage fails and the lizard is detected, it runs as best as a small pancake can. Often this amounts to sprinting clumsily for a short distance and then stopping to freeze again. You'd be surprised how easy it is to lose track of a horned lizard trying to escape; ask my husband.

When an observant predator corners a horned lizard, the lizard might raise up high on all four legs, lower its head, and charge like a mad bull. Or it might puff up its body by inflating its lungs, thereby making itself too broad, spiny, and plump for the predator to swallow. Often it opens its mouth and hisses ferociously. Usually this is a bluff, but occasionally it bites. A horned lizard might vibrate its tail, which if done on dry leaves sounds frighteningly similar to a rattlesnake's buzz. Sometimes the lizard will arch its body and rock forward and backward on its extended legs. When you pick one up, it might struggle and jab its horns into your fingers. If attacked, it might bend its head down and arch its back, which raises the profile of the horns and makes it harder for a predator to swallow it.

On the other hand, a hand-held lizard often becomes immobile, as if hypnotized, especially if you stroke it between the eyes or turn it over and rub its belly. Perhaps this "death-feigning" behavior deters predators that dislike eating dead animals.

As per the quote, some horned lizards also squirt narrow streams of blood from their eyes at predators. The lizard arches its back and closes its eyes. The eyeballs protrude and the eyelids engorge with blood until the increased blood pressure in the eye-socket sinuses causes the walls to burst. A fine stream of blood shoots out from either one or both eyes, up to four feet either frontward or backward.

Although this blood-squirting behavior has been reported since the days when the Spaniards first explored Mexico, we're only beginning to learn something about the circumstances under which it occurs. In the 1960s, investigators noticed that horned lizards squirt blood when molested by members of the dog family. Coyotes, foxes, and domesticated dogs shake their heads and salivate excessively when a horned lizard sprays blood into their mouths or nostrils, presumably because the blood is distasteful to them. In contrast, people who frequently handle horned lizards report that the animals only rarely squirt blood onto them, and the blood doesn't taste objectionable.

Are horned lizards really more likely to squirt blood at dogs than at people who harass them? This question was answered by a simple but effective experiment. During 1989–90, Wade Sherbrooke, director of the Southwestern Research Station in Portal, Arizona, and a colleague experimented with Texas horned lizards and showed that indeed these lizards more readily squirt blood at dogs than at humans. In each of ten trials, they placed one lizard in a sandy, open arena with Dusty, a yellow Labrador retriever. In another ten trials, they placed a lizard in the arena with Wade, who pretended to be a dog. Dusty attacked all ten lizards repeatedly—barking, pawing, nipping, and picking up and tossing the lizard. In all ten trials, the lizards squirted blood. In the other ten trials, Wade got on all fours, barked at, chased, and harassed the lizards. In only two of these trials did the lizards squirt blood. The lizards used other means of defending themselves against Wade, including openmouthed charges.

AND SO IT seems that horned lizards distinguish between canine and human predators, displaying defensive behavior appropriate to each. At least in some parts of North America, however, this unique defense of blood squirting may increase the lizards' social standing. Many Mexicans revere and protect horned lizards because they believe the lizards to be sacred: their tears of blood forge a connection with Christ. In an odd twist of fate, then, blood squirting works as a defense against humans after all.

CASTING THE INSIDES OUTSIDE

The Sea-Cucumber Foils the Fish
(*Apologies to Longfellow*)

The hungry fish attacked Thyone
Who gave himself, himself alone.
Bestirred anon his insides inside,
Turned at length the insides outside.
To get the en'my outside offside,
Turned his viscera inside outside.
Pushed his inside topside outside
Till the fish's hunger capsized.
Tuck't remainder outside inside,
And grew more insides in the inside.

BERNAL R. WEIMER, from *Nature Smiles in Verse*

What sense does this seemingly nonsense poem make? Read it again. The poem describes the ingenious way a sea cucumber outwits a predator and manages to stay alive. When a hungry fish attacks a sea cucumber of the genus *Thyone*, the cucumber contracts its muscles and ejects its intestines. The fish eats until satisfied, then the cucumber gathers back the remainder of its insides. Still alive, the somewhat mangled animal regenerates new body parts over time.

Have you ever seen a sea cucumber up close and personal? Ever held one? It's difficult to tell the front end from the back end if the ten to thirty feeding tentacles surrounding the mouth are pulled in close to the body. If that's the case, all you have in your hand is

a large flaccid pickle. Relatives of starfish and sea urchins, sea cucumbers are marine creatures shaped like their namesake vegetables. They're more variable in color than the vegetable, however: reddish brown, yellowish brown, tan, black, or nearly white. These sluggish animals, the consistency of a soft leather pouch, live on the ocean bottom or on rocky shores, where they attach themselves to rock ledges and hide in crevices. Their soft bodies and fairly sedentary nature make them vulnerable to hungry fish, crabs, and other predators. Hence, the effective defense.

Different species of sea cucumbers provide this decoy in different ways. Some blow their intestines out through the front end. Others blow them out through the back end. In some the body wall breaks apart, spilling intestines from the middle. Sometimes the animal tosses out its respiratory organs and gonads along with its digestive tract.

Some sea cucumbers also drop off part of their bodies when attacked—an act of self-amputation called autotomy. The remaining tissues get redistributed and remodeled, eventually producing a completely functional animal. Even if the loss includes more than 95 percent of the animal's digestive tract, as long as the most anterior portion is intact, a new animal will develop inside the leathery skin.

Certain sea cucumbers combine autotomy with another trick. When threatened by a crab or other predator, the cucumber aims its anus at the attacker and contracts its body wall, which ruptures the hindgut and discharges sticky tubules associated with its digestive system. The tubules swell into adhesive threads, entangling the attacker in a sticky mass of goo. The sea cucumber can regenerate the tubules and any other tissue lost during its dramatic attempt to save itself, provided the defense works and it's not ingested whole by the predator. But what a hernia!

ODD AS SEA cucumber defense appears, other animals can voluntarily self-amputate parts of their bodies in response to predator attack and then regenerate the missing part. One example is an animal many of us consider a delicacy: the lobster. A lobster has five pairs of walking legs, the front pair modified as huge claws filled with powerful (and tasty) muscles. If a predator grabs and holds on to any one of these legs, including the claws, the lobster may drop it off and scurry away. If the detached body part is a claw, it may grasp the attacker even more strongly than if it were still attached to the lobster. If a nonpincher leg drops off, the lobster can still scoot away while the predator consumes its smaller-than-expected meal. In time the lobster will regenerate the lost leg. A crab can do the same. In some populations of blue crabs, nearly 40 percent of individuals have missing legs.

Starfish sacrifice their arms to predators as an escape tactic. Most starfish have five arms, though some have forty or more. The arms allow a starfish to move around in its environment, but in addition each arm houses a nerve cord and a pair of digestive glands. At the end of each arm, a light-sensitive eyespot distinguishes light from darkness. In most species, the sexes are separate, so each arm also houses either its own egg-producing or sperm-producing organs. As important as the arms are, a starfish will drop one or more when attacked. It can always grow more. Even if only one arm is left, all the missing ones will usually regenerate—as long as that one arm is still attached to a significant portion of the central disk. There is one exception, however. Starfish of the genus *Linckia* can regenerate an entire new body from a detached arm isolated from its central disk.

A starfish opens its favorite food—oysters, mussels, and clams—by spreading its arms over the two valves (shells). Then it pulls the shells in opposite directions. Certain species of starfish move about on the ocean bottom in hordes, eating every shellfish within reach. Much to the dismay of fishermen, these starfish can demolish an oyster bed almost overnight. Even low densities of starfish can wreak havoc in a fisherman's favorite shellfish-gathering area. Many a fisherman naive of a starfish's ability to regenerate its arms has ripped apart starfish in anger and dumped the pieces into the ocean. Little did they realize that if any of those arms were still attached to a significant portion of the central body region, it would become a complete animal. What was one pest in time could become several pests. And instead of one reproducing animal, in time there could be several.

Brittle stars, agile relatives of the sluggish starfish, voluntarily drop their arms when attacked by predators. Some can even cast off the central part of the body, discarding the stomach and other tissues. The missing arms grow back, and even the central portion and associated organs regenerate.

Let's move onto land. The long legs of some centipedes protect them from attack since a predator has trouble grasping a centipede's body with its gangly legs flailing every which way. Unable to grab the body, an attacker often goes for a leg, which the centipede quickly autotomizes. The cast-off leg continues to wiggle, distracting the attacker and allowing the centipede to escape. A drop of yellowish viscous fluid is exuded at the point where the leg was attached. The drop hardens and prevents loss of body fluids, and eventually the centipede regenerates a new leg. When attacked, some centipedes heighten the drama and confuse or frighten the predator by swinging their hind legs from side to side and making a loud creaking noise by rubbing tiny hooks on one part of the leg against another part of the leg.

Some vertebrates also sacrifice body parts as an escape tactic, then regenerate the lost parts later. The classic example is tail loss in lizards and salamanders. Birds and mammals often attack the head end of their prey. A lizard or salamander that wiggles its tail diverts the predator's attention to the less vulnerable end of its body. Then, if attacked, the lizard or salamander simply drops part of its tail. The still-wiggling appendage continues to hold the predator's attention, and the lizard or salamander escapes while the predator snacks on its tail.

In lizards and salamanders that autotomize their tails, the tail breaks off easily at fracture planes located in the middle of certain vertebrae or between certain vertebrae. The particular arrangement of muscle and connective tissue in the tail also allows for easy separation. In some species, there is only a single fracture plane, so the tail can break off only in one place. In most species, however, there are several fracture planes along the tail length. Although most lizards and salamanders regenerate the tail after breakage, the new tail is shorter than the original one and it may be a different color. A regenerated tail is supported by a solid rod of cartilage instead of bony vertebrae, so it can't be autotomized again.

Obviously, if a lobster, crab, starfish, or centipede is faced with losing either an appendage or its life, its better bet is to drop the limb. But autotomy comes with a cost. Growth may be retarded because energy is diverted to growing the new body part. The animal may be less efficient at foraging for food, have a harder time escaping from predators, and be less likely to win fights. Limb loss might also reduce an animal's reproductive success by decreasing its ability to attract or keep mates or by physically hindering mating. All of these costs also apply to lizards and salamanders that autotomize their tails, but these vertebrates incur an additional cost. Because fat is stored in the tail, loss of the tail may prevent the animal from breeding if it doesn't have enough reserve energy; or even if it breeds, it might produce fewer offspring than if it still had its tail.

WHEN ATTACKED, BIRDS and mammals can't drop body parts and then regrow them. The best they can do is regenerate bone, muscle, skin, and nerve fibers that have been injured. It's rather humbling to realize that although humans are supposedly the "highest" of vertebrates, we are impotent in some areas. We can regenerate fingernails and hair, but once an arm or leg is lost . . . it's gone. The reason is that the cells left behind at the base of the severed human arm or leg can't develop into new tissues. In contrast, when a lobster, lizard, or the like autotomizes a body part, the cells left behind activate and organize into the appropriate new tissue.

Even if humans can't regenerate lost limbs, exciting research suggests that we may be able to repair cells, tissues, and even organs by introducing "stem cells" into a diseased or injured person. Stem cells are precursor cells that give rise to tissues. Because many diseases are associated with the death or dysfunction of a single type of cell, if we could replace these cells with healthy ones, perhaps we could repair tissues that don't normally repair themselves. For example, our bodies can't regenerate damaged heart muscle. But if we could culture stem cells in the laboratory and make them into cardiac muscle cells, we could inject them into a damaged heart with the hope of repairing the muscle tissue. Someday persons with Alzheimer's disease, Parkinson's disease, autoimmune diseases, osteoporosis, juvenile diabetes, cancers, and spinal cord injuries might also benefit from stem cell therapy.

Where do stem cells come from? We all have them: embryos, children, and adults. Embryonic stem cells are the most useful because they can differentiate into virtually any type of cell. But there's a negative side to embryonic stem cells. Even if the embryo comes from an egg fertilized in a petri dish, it is still a potential future human being. And therein lies the problem. Ethical, legal, and regulatory considerations come into play because extraction of stem cells kills the embryo. Stem cells from children and adults might provide a viable alternative, but they're less versatile: the cells generally differentiate into cells only from the same type of tissue. For example, stem cells from bone marrow differentiate into bone marrow cells, but not into muscle tissue cells.

ADMIRE THE LOBSTER, crab, starfish, centipede, salamander, and lizard that can drop a body part and then regenerate a new one. And admire the sluggish sea cucumber that can turn its insides outside and then grow more insides inside. Regardless of stem cell advances, we'll never have the natural regenerative powers of certain invertebrates and "lower" vertebrates.

SPIT 'N' SPRAY

The foul smell of the cud is just as obnoxious for the spitter as the spittee. Therefore llamas spit only as a last resort. You can tell when two llamas have had a disagreement—they both walk around with their mouths open, trying to get rid of the terrible odor.

DAVID HARMON and AMY S. RUBIN, *Llamas on the Trail*

"Run!" I yelled to my seven-year-old daughter, who was already sprinting as fast as she could away from the overprotective mama llama chasing her  at full speed, spitting white foamy saliva. I was stuck in the car, with my foot on the gas pedal, because our ancient Peugeot had ignition problems and we hoped to avoid being stranded in the high Andes west of Tucumán, Argentina. My husband was photographing llamas some distance away while Karen had been strolling in the grass, approaching a fluffy brown-and-white baby llama. Karen raced to the car and dove onto the seat just ahead of the angry mama llama.

I turned back to console Karen and felt something nudge my shoulder. Whipping around to the front, I stared straight into the mama llama's face, her vomit-smelling, spit-covered nozzle about two and a half inches from my open mouth. It seemed an eternity between the time I hit the button for the power window and the time her head was safely outside the car. As a parting gift, she smeared saliva on the glass.

The smell of llama spit ranges from really bad foot odor to rotting garbage to vomit. Why so bad? Like cows, llamas are ruminants. While grazing, a llama doesn't chew its food well before swallowing. The food goes into the first chamber of its stomach, where it's stored until it softens into a glob called the cud. While the llama is resting, muscles in this chamber send the cud back up into its mouth. The llama now chews its cud more thoroughly before swallowing, at which point the food heads back down into other chambers where digestion continues. If you chewed regurgitated food, your saliva might reek too!

Spitting foul-smelling saliva is one way llamas and some other members of the camel family defend themselves, but they aren't the only spitters. And some animals spit more than stenchy saliva. Some, such as spitting cobras, spit eye-blinding toxins in self-defense.

Cobras, which live in southern Asia and Africa, have potent venom and fangs in the front of the upper jaw. Some species inject the venom by sink-

ing their fangs into another animal's flesh. In these species, the openings
in the fangs through which the venom is ejected point downward. Others
have fang openings that point forward, allowing them to spit venom at
their attackers. Spitters rear into an upright posture, spread their hoods,
tilt back their heads, and aim twin jets of venom at their adversaries' eyes.
They can accurately spit droplets of venom at targets up to eight feet away,
causing eye pain, irritation, and even permanent blindness.

South Africa's most notable serpent spitter is the ringneck spitting
cobra. Fortunately for humans, this snake prefers to spit rather than bite.
A bite to a person can result in unconsciousness within two minutes and
respiratory paralysis, requiring antivenin. In contrast, spit in the eye causes
inflammation, pain, and extreme sensitivity of the eyes to light, but the
symptoms usually disappear within two or three days. Instead of antivenin, victims need only wash their eyes thoroughly with water.

Spiders in one family, the scytodids, are called "spitting spiders" because they spit a mixture of venom and glue to immobilize their prey. Like
many animals, these spiders also use their secretion for another purpose:
defense. They spit their concoction at close range onto a scorpion or other
predator. While the attacker tries to untangle its legs, the spider escapes.

In the early 1980s Linda Fink, a friend who was studying female green
lynx spiders guarding their egg sacs, discovered that these spiders (unrelated to scytodids) also spit venom. While censusing the spiders, Linda occasionally noticed droplets of fluid on her face and hands: venom from the
females' fangs. The droplets irritated her eyes. If Linda gently tugged a female's front leg, the spider would face her, lunge slightly forward, and spit
droplets that landed up to eight inches away. Linda concluded that the
venom may be an effective defense against egg predators such as ants, or
it might protect the female herself.

Instead of spewing from the mouth, some animals spray their toxins or
offensive fluids from other areas of the body. If you've never had a run-in
with a skunk, consider yourself fortunate. When frightened or threatened,
a skunk offers a characteristic three-part warning: it stamps its front feet,
lifts its tail, and hisses or growls. If its warn-
ing is ignored, it blasts the intruder with a
yellowish fluid from a pair of glands located
just inside the anal tract near the base of
the tail. This fluid, known to chemists as
n-butyl mercaptan, contains sulfur. The
sulfur adds to the foul odor. A skunk can
spray accurately up to fifteen feet, and it

can fire up to six rounds in rapid succession. After firing all its ammunition, though, the skunk needs several days to produce enough mercaptan to fully reload. For this reason, it doesn't spray unless absolutely necessary. Thus, the warning. Skunk odor lingers for days and can be detected up to half a mile away. The spray can temporarily blind a person or another animal. Baby skunks become little stinkers at about seven weeks, fully capable of spraying in self-defense.

So what can you do if a skunk has sprayed you or your dog? Some people swear by tomato juice baths, but it takes a lot of tomato juice to make you or your dog presentable. My dogs' veterinarian claims this antidote is just an old wives' tale anyway. While in vet school, he learned about an alkaline hydrogen peroxide solution, which he swears by: one quart of 3 percent hydrogen peroxide mixed with one-fourth cup of baking soda and one tablespoon of liquid soap—good for either clothes or fur. Another vet at the same clinic recommends a 2 percent vinegar solution or a Massengill douche, which is basically vinegar and water. Commercial products that break down the oily components of skunk spray work also: Odor Breaker, Krystal Air, and Nature's Miracle Skunk Odor Remover.

Humans have figured out how to exploit skunks' defensive spray. Authors Lauren Livo, Glenn McGlathery, and Norma Livo write in their book *Of Bugs and Beasts:* "Rocky Mountain Wildlife Enterprises sells a skunk-based product called Scrooge Christmas Tree Protector. Park managers and others spray the tree protector on pine trees or other shrubs that might be tempting targets for Christmas tree poachers. The smell discourages would-be tree thieves." No doubt we can all think of other possible uses for the commercial spray . . .

The European fire salamander sprays toxin from its back. A double row of large skin glands sits along the upper side of the salamander's boldly colored black-and-yellow body. These glands contain a potent toxin that affects the central nervous system and can cause respiratory paralysis. When molested, a fire salamander extends its legs, elevating and tilting its body toward the attacker. It contracts the muscles around the glands and sprays their contents up to six feet—impressive for an animal that, including the tail, is only about ten inches long.

Insects are experts at chemical warfare. (For a very readable account of insect chemical warfare, I recommend Thomas Eisner's 2003 book *For Love of Insects.*) Some of the better-known spraying insects include certain members of the following groups: cockroaches, walkingsticks, European earwigs, assassin bugs, European wasps, and beetles. Some ants bite their attackers and then squirt formic acid into the newly inflicted wounds.

Anteaters specialize on ants and termites, but they have to deal with the nasty defenses of those insects. In some species of termites, the soldiers (the ones that rush out of the nest when it's broken into) have long snouts that work as spray guns. They aim at the anteater and spray a sticky repellent that smells (to humans) like pine oil. Many ants also spray strong-smelling substances to defend their nests from anteaters and other predators.

While taking students on a nature walk one evening in montane rain forest of southwestern Colombia, I found a pair of copulating walkingsticks slowly swaying in the breeze, mimicking twigs. The female was nearly a foot long; her piggyback riding companion was barely a third her size. I explained to the students that these insects were using camouflage to avoid being detected, but that they had another trick if that didn't work. I picked up the pair and sure enough, the couple responded on cue and discharged a foul spray onto my hands—foul enough that I quickly replaced the insects on the branch while the students snickered.

You needn't go to the tropics to get sprayed by a walkingstick. If you touch the six-inch southern walkingstick (*Anisomorpha buprestoides*), common in Florida, it will spray. Southern walkingsticks are especially aggressive toward birds. A bird need only approach, and the insect sprays. The irritating secretion is produced and stored in two glands in the thorax, opening just behind the head. The considerably smaller male spends most of his life riding piggyback on a female, even when not mating. Could it be that pairs, capable of discharging a double whammy, can defend themselves more effectively than can solitary individuals?

In the mid-1960s Thomas Eisner, an ecologist from Cornell University, wondered how the flightless, slow-moving jet-black beetle *Eleodes longicollis* (family Tenebrionidae) can survive the Arizona desert predators abundant at dawn and dusk: ants, centipedes, spiders, scorpions, toads, and rodents. When he gently tapped or prodded these beetles, they stopped abruptly, raised their rears into the air, effectively doing a headstand, and sprayed a smelly golden-brown liquid. The liquid turned Eisner's fingers purple—a transformation that lasted for days, until the skin wore off.

The spray of these beetles, appropriately called bombardier beetles, comes from a pair of glands that open at the tip of the abdomen. Each gland consists of an inner and an outer compartment, each of which contains different chemicals. To discharge, a beetle squeezes hydroquinones and hydrogen peroxide from the inner compartment into the outer compartment, where enzymes trigger an instant chemical explosion. The hydroquinones get converted to nasty-smelling quinones, and the hydrogen peroxide breaks down to oxygen and water. The pressure of the oxygen gas causes the mixture to blast out of the beetle's hind end. As a result of these chemical reactions, the temperature of the spray is about 212°F, the temperature of boiling water and hot enough to have a thermal as well as an irritant impact.

The bombardier beetle's defense is so effective that other beetles cash in on the idea. Another jet-black tenebrionid beetle (*Megasida obliterata*) that roams the desert alongside *Eleodes* also comes to a sudden halt when disturbed and promptly does a headstand, lifting its rear upward. Eisner found, however, that this beetle lacks stink glands and doesn't spray. It bluffs its potential predators by mimicking *Eleodes.*

Real bombardier beetles can spray in any direction, and they aim accurately. Their spray is an effective defense against most predators, but there are exceptions. Toads are quicker than bombardiers. As soon as a beetle is within range, a toad flips out its sticky tongue and then retracts the tongue and attached beetle into its mouth before the victim can spray. A grasshopper mouse grabs a beetle with its front feet and jams its rear end into the soil or sand. The mouse contentedly munches away while the beetle frantically discharges its stench into the ground. All that remains after a meal are the beetle's legs, front wings that are fused into a protective shield, and the tip of the abdomen containing the offensive glands. Orb-weaving spiders deal with these beetles in yet another way. When a beetle lands in an orb-weaving spider's web, the spider carefully wraps the intruder in silk so gently that the beetle doesn't spray. When the spider sinks its fangs into its victim, the beetle sprays, but the hot irritant remains trapped in the silk.

Moving from the Arizona desert to the moist tropical rain forest . . . other sprayers are the onychophorans. These soft-bodied terrestrial animals with large antennae and many stumpy legs belong to a small

invertebrate phylum consisting of about eighty species. They're often called "missing links" between annelid worms (segmented worms such as earthworms, polychaetes, and leeches) and arthropods (invertebrates with exoskeletons and jointed appendages, such as insects, spiders, and lobsters) because they share characteristics of both phyla. Like annelids, onychophorans have unjointed appendages; their excretory system is also annelid-like. Like arthropods, they shed their exoskeleton; their circulatory and respiratory systems are arthropod-like.

These unusual creatures capture their food in a unique way. Once an onychophoran detects dinner—a termite or other insect, snail, or small worm—it discharges two powerful streams of glue from slime glands that open on the head. *Zap!* The adhesive slime hardens quickly and entangles the prey in a weblike net.

Like the spitting spiders, onychophorans also use their secretion for defense. They can spray their slime a foot or more to protect themselves from centipedes, spiders, ants, and other predators. While the predator struggles to free itself, the onychophoran escapes. Sometimes the would-be predator can't disentangle itself, so it dies glued in place.

AS WE'VE SEEN with the onychophorans, bombardier beetles, and certain other animals, once a predator has detected and cornered a prey animal, an effective defense may be spitting or spraying an offensive or toxic fluid to repulse the predator and avoid getting injured. Theoretically a great strategy, it works against many predators but not all. Recall the grasshopper mouse's strategy for eating bombardier beetles: jam the offending end into the ground. And the orb-weaving spider's strategy: wrap it in silk to contain the stink.

Predators and prey engage in an evolutionary arms race. As prey develop more effective ways of avoiding or outsmarting predators, predators develop more effective ways of breaking through their prey's defenses.

VENOM: SERPENTS' FANGS

With thy sharp teeth this knot intrinsicate
Of life at once untie. Poor venomous fool,
Be angry, and dispatch.

SHAKESPEARE, *Antony and Cleopatra*

So implored Queen Cleopatra to the asp (probably an Egyptian cobra) as she pressed it to her breast. Mark Antony had stabbed himself, thinking Cleopatra was dead. Imprisoned, Cleopatra used the serpent to carry out her suicide. Cleopatra is undoubtedly history's most famous suicide by snakebite, but more than one well-known herpetologist has accidentally lost his life to a venomous snake.

Herpetologists thrill in sharing their tales of venomous snakebite. Several close friends of mine have been bitten; fortunately, all have lived to tell their stories. I've never been bitten, but I've had several close encounters. My most unsettling encounter with a venomous snake occurred in the rain

forest of eastern Ecuador, when I was twenty-five years old. I was returning from fieldwork at about 2 : 00 a.m. The moon was shining brightly, so
I had turned off my flashlight to save on batteries. After about twenty minutes, for some unknown reason I turned on my light again. Less than three
yards ahead, a seven-foot bushmaster viper faced me in a coiled ready-to-
strike position. The black blotches along its back seemed poised to leap off
its tensed pale brown body. I stumbled to a halt. My heart pounded furi-
ously. Finally, I took a wide detour around the snake. The bushmaster's
scientific name, Lachesis muta, translates into "silent fate," an appropriate
name for a venomous snake with long fangs capable of injecting a massive
dose of venom.

Recent polls in the United States reveal that people fear snakes more
than spiders, mice, or even public speaking. Nevertheless, of the nearly
2,700 species of snakes, only about 270 have venom potent enough to
harm or kill a person. Venomous snakebites in the United States average
1,000 to 2,000 per year, causing 10 to 15 deaths. Let's put this into perspective. An average of 42,271 people in the United States die every year
from car accidents, 14,053 from unintentional poisoning, and 3,179 from
drowning. Elsewhere in the world, especially in the tropics, venomous
snakebite is a serious hazard because many rural people go barefoot,
proper medical attention is difficult to find, and the tropics harbor more
venomous snakes. Estimates of annual worldwide venomous snakebites
range as high as 1 million, of which up to 40,000 are fatal.

A commonly asked question about venomous snakes is: Which one is
the most dangerous to humans? This is a hard question to answer, because
it depends on many factors. From the standpoint of potency, drop for
drop the venom of the Australian inland taipan is probably the most toxic.
A bite from one of these eleven-foot giants contains enough venom to kill
200,000 mice or certainly several people. Fortunately, these snakes are
only spottily distributed in northern Australia, and they would rather retreat than bite. Because they occur in areas of low human population den-
sity, contact with people is rare. Other snakes are less venomous drop for
drop but are more likely to live around people. Some of the contenders for
snakes that kill the most people include Indian cobras from southern Asia;
tiger snakes from Australia; black mambas, puff adders, and saw-scaled vipers from Africa; and fer-de-lance from Central and South America. None
of the top killers occurs in the United States. Recall that we have only ten
to fifteen instances of fatal snakebite per year.

The main function of a snake's venom is not to bite people. Venomous
snakes use venom and a venom delivery system to subdue large struggling

prey. Their injection mechanism works like a hypodermic syringe. Venom glands serve as a syringe barrel for storage. Jaw musculature functions as a syringe plunger. And fangs deliver the liquid just as a needle does. Although the primary function of venom is prey capture, when a venomous snake is threatened, it may bite to defend itself.

The three main groups of venomous snakes are the vipers, including rattlesnakes, copperheads, cottonmouths, puff adders, and horned vipers; the elapids, including cobras, mambas, coral snakes, and sea snakes; and rear-fanged snakes, including African boomslangs, savanna twigsnakes, Blanding's treesnakes, and South American green racers. These three groups are distinguished by differences in their fangs. The significance of these differences for you and me is how the snake bites.

Vipers have long movable fangs in the front of the upper jaw. Venom flows through these tubular fangs into the wound. When not in use, the fangs fold back and lie against the roof of the mouth. These fangs are the most advanced and efficient of the three types, allowing vipers to strike rapidly, inject the venom deeply, and then quickly release the prey or attacker. The entire strike and envenomation sequence of a rattlesnake, for example, lasts about 200 milliseconds. Let's take a closer look at rattlesnakes— thirty-three species that, as a group, occur from Canada to Argentina.

Rattlesnake bites account for most of the fatalities from snakebite in the United States. The most deadly rattlesnakes to humans are the eastern and western diamondbacks. Both can be cranky and quick to strike. Associated with each fully functional fang, rattlesnakes have up to six additional fangs in various stages of development. The spare fangs are critical because a rattlesnake often swallows its functional fangs along with its prey. When a fang breaks off or is lost by periodic shedding, a new one stands poised for action. On average, a functioning fang lasts only about six to ten weeks.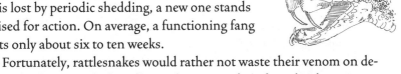

Fortunately, rattlesnakes would rather not waste their venom on defense, so they warn before they strike. Except for a few island species, rattlesnakes have loosely connected, interlocking horny segments at the ends of their tails: their rattles. By vibrating their tails, the snakes create a distinct buzzing sound that warns of their presence. After a snake sheds its skin, it adds another rattle segment. A popular myth is that you can tell a rattlesnake's age by the number of segments. Not so. A snake may shed two or three times each year. Furthermore, it's rare to find a rattlesnake with more than nine segments. Old segments break off from wear and tear.

For a while our forefathers considered making the rattlesnake our national emblem. Rattlesnakes figured prominently on colonial flags: often in a coiled ready-to-strike position with the motto "DON'T TREAD ON ME." During the American Revolutionary War (1775–83), the Continental navy also flew a red-and-white thirteen-striped flag with a rattlesnake design and the words "DON'T TREAD ON ME." Rattlesnakes were chosen as a symbol during this period in American history for several reasons. They represented vigilance, deadly striking power, and ethics, since they warn before striking. Rattlesnakes don't initiate fights, but once engaged they don't back down. Likewise, the colonists had no intention of surrendering their independence.

The second group of venomous snakes, the elapids, have short, immobile (or only slightly mobile), permanently erected fangs on the front of the upper jaw. Their fangs are either tubular or deeply grooved. These snakes typically hang on to their victims and chew venom into the wound.

The only elapids in the United States are the yellow-bellied sea snake off the coast of California and two species of coral snakes—one in central Arizona and southwestern New Mexico and one in the east from North Carolina to west-central Texas. As a kid, I learned the ditty: "Red on black, venom lack; red on yellow, kill a fellow." This works to distinguish the two black-red-and-yellow-ringed coral snakes from their nonvenomous mimics in the United States, but *not* in the tropics, where any combination is possible. Both of the U.S. coral snakes have red and yellow rings touching each other; the mimics—such as scarlet snakes, scarlet kingsnakes, and several species of milk snakes—have a black ring in between each red and yellow ring.

Cobras are common elapids in some parts of Asia. After reading Rudyard Kipling's "Rikki-Tikki-Tavi" to me, my father shared his experiences with cobras in the Philippines. He often saw them along the jungle trails on Mindanao. He didn't bother them, and they didn't bother him. Once he was sitting on his nipa hut porch with six or seven Filipinos, all in

shorts and barefoot, when a trio of thirty-inch cobras crossed the bamboo floor in single file. One of the Filipinos quietly said, "Don't anyone move." No one did. The cobras nonchalantly crossed the porch, slithered between two of the men, and disappeared down through the slots.

Snake charmers from Asia and northern Africa play their flutes to cobras, rather than some other type of serpent, because cobras are impressive beasts. They're big. And they're dangerous. When excited, cobras rear upright, spread their hoods by expanding their ribs, and sway. The snakes can't hear music; they're simply holding themselves on guard, swaying in response to the flutist's body movements. Fortunately for snake charmers, the snakes' strikes are relatively slow and usually can be evaded.

The third group, rear-fanged snakes, have one to three enlarged teeth in the rear of the upper jaw. Venom flows down grooves in the teeth into the bite wound. Most rear-fanged snakes are not considered dangerous to people, because to get enough venom into you the snake really has to chew—and you're not likely to allow it to do that. There are exceptions, however. An African boomslang killed the prominent herpetologist Karl P. Schmidt in 1957 as he handled it a bit too fearlessly. In 1972 a savanna twigsnake killed the renowned German herpetologist Robert Mertens as he hand-fed his pet. Bites from other rear-fanged snakes have caused severe headache, nausea, pain, and even death to people who handled the snakes not realizing the potential danger.

Snake venoms destroy specific targets, such as blood or nerves. In general, a viper bite causes immediate pain, swelling, discoloration, weakness, and nausea. An elapid bite may not hurt at first, but eventually the victim becomes sleepy and has trouble breathing. These generalizations are misleading, however, as many snake venoms both destroy tissues and cause respiratory distress.

People have long used snake venom as medicine. In the twelfth century, physicians treated leprosy and cancer with snake venom. During the first half of the 1900s, many doctors considered venom from rattlesnakes, cobras, coral snakes, and certain tropical pit vipers to be a valuable drug. They used it to treat dozens of conditions and diseases, including hemorrhaging, epilepsy, and severe pain from malignant tumors. Physicians usually administered the venom by mouth in dilute form but sometimes injected it or made it into an ointment or wet compress. For the past half century, most doctors have believed these conditions are more safely treated by conventional medicines. The pendulum is shifting in the other direction, though. Current medical research suggests that someday snake venom may be used to selectively destroy certain types of tumors, treat osteoporosis, lower blood pressure, promote blood clotting, reduce pain, and combat bacterial infection.

Antivenin serum is currently the most common use of snake venom. Experts extract or "milk" snakes of their venom by grasping the snake behind the jaws, and when the snake opens its mouth, they hook the fangs over the edge of a collecting vessel. The angry snake squirts venom. The vessel is usually covered with a membrane so the snake's mouth secretions won't contaminate the venom. Technicians freeze the liquid venom, dry it under strong vacuum, and then make antivenin by injecting venom into horses or sheep in increasing sublethal doses until the animal achieves immunity. They draw blood from the animal and purify the serum containing antibodies against the venom. When this purified serum is injected into a snakebite victim, the venom is neutralized. Venom extraction facilities operate around the world to provide antidotes, so that if you or I get bitten by a venomous snake, we may survive.

The Irula Snake Catchers' Industrial Cooperative is India's largest producer of snake venom, and its conception is an unusual story. Between the 1930s and 1970s, many of the rural Irula tribal people of southern India earned a living by catching snakes and selling them to tanneries. During a boom period between the early 1950s and the 1970s, India exported an average of 5 million skins each year—skins destined to become luxury belts, shoes, and handbags. Many species became rare. Rats, a main food of snakes, became so abundant that they destroyed 30 to 50 percent of the annual grain crops. Finally, in 1976 the India government banned exportation of snakeskins—legislation that was good for snakes and farmers, but bad for snake catchers.

As an alternative for the Irula, a cooperative for venom extraction was formed. Members of the cooperative catch cobras, vipers, and kraits; extract their venom three times each; and then return the snakes to the wild. Everyone wins: the snakes live and continue to hunt the rats that eat the grain, India's supply of antivenin serum has been increased, and the Irula earn an income many times greater than what they earned selling snakeskins. Skins from fifty kraits are worth only about three hundred rupees. Venom milked from fifty kraits pays about ten times that amount. Worth five to a thousand times as much as gold by weight, snake venom is one of the world's most valuable natural resources.

BUT GETTING BACK to the world's most famous suicide by snakebite victim . . . Cleopatra lived fast, died young, and chose her weapon wisely. Envenomation by a viper would not have left a good-looking corpse, but rather one swollen and disfigured. Envenomation by a cobra left her more dignified. She likely had difficulty breathing, but felt little pain or anxiety, lapsed into a coma, and died peacefully and beautifully—a calculating woman to the end.

VENOM: CENTIPEDE AND SPIDER FANGS

The Centipede

A centipede was happy, quite,
Until a toad in fun
Said, "Pray, which leg moves after which?"
Which raised her doubts to such a pitch,
She fell exhausted in the ditch,
Not knowing how to run.

ANONYMOUS, from *Nature Smiles in Verse*

As the title suggests, centipedes have fangs and venom. But before we go
there, let's follow up on the confused centipede just described. After all,
a centipede's first line of defense against predators is simple: just run.

How do centipedes manage to run without tripping over their own legs? To begin with, a centipede's
body hangs down from its legs, rather than sitting up
above vertical legs as in the case of most mammals.
This provides a low center of gravity and stability. Also,
because a centipede's body is jointed, each segment can
bend back and forth, making its body far more limber than yours
or mine. The legs of many species are progressively longer toward
the back end of the animal, helping the centipede to push off to a running
start. When a centipede runs, its legs move rhythmically in waves, first on
one side of the body and then the other. Thus, at any given moment, the
legs will be bunched up on one side and spread out on the other. Even
though the tips of the legs cross over each other when the animal is running, the centipede doesn't trip. By the way, it's a myth that all centipedes
have a hundred legs. Some have as few as 30 legs; others have as many
as 180.

As a second line of defense, some centipedes secrete foul-smelling,
caustic, or slimy substances. Some centipedes rapidly and vigorously twist
their bodies and make themselves hard to catch and hold. And some purposely self-amputate a leg held by a predator.

And now to those fangs. Centipedes, those multi-legged wonders scuttling along your bathroom floor, look sinister for a reason: the first pair of
legs is modified into fangs. A gland at the base of the jaw supplies venom,
which the centipede injects into its prey: spiders, insects, worms, and even
small amphibians, reptiles, birds, and mammals, depending on the cen-

tipede's size. Some centipedes can rear up on their back segments and catch bees and wasps flying by. After injecting venom, they drop the prey and wait until it's paralyzed before eating it.

Centipedes also bite to protect themselves, as when you try to remove them from your bathtub. Although the venom of most centipedes isn't dangerous to people, some tropical species are exceptions. If you visit India or Burma, watch out for their foot-long centipedes, whose bites can keep you bedridden for two or three months. What a way to spoil a vacation! We have a huge centipede in southern Arizona: the giant desert centipede, an eleven-inch monster that eats small lizards. Although the mere sight of this centipede is terrifying (especially as it strolls out from your sleeping bag) and the bite is painful, generally the venom is not life-threatening to people.

Centipede biologists warn nonspecialists to treat all of the more than fifteen hundred species of centipedes with respect: if you're not an expert at identifying centipedes, leave them alone. Their bites can cause intense local pain, vomiting, headache, swelling, numbness, and necrosis (tissue death). Instead of disturbing them, just appreciate them for the unique wonders they are: multi-legged creatures that have evolved a complex arsenal of defenses to protect themselves.

ALL SPIDERS HAVE fangs at the tip of a pair of appendages above the mouth, just below the eyes, and most have venom glands that service their fangs. A spider stabs its prey with its fangs, delivering venom that either kills or paralyzes the victim. Some spiders also inject venom to defend themselves.

The female black widow spider—a handsome, shiny black spider with a red hourglass-shaped mark on the underside of her abdomen—is the most dangerous spider in the United States and perhaps in the world. Her venom is fifteen times more toxic than that of a prairie rattlesnake—impressive

since her abdomen is the size of a pea. Females are shy and usually try to escape if disturbed, but they'll bite if you touch them or if they get trapped beneath your clothes. The males are smaller, and they don't bite. In fact, their venom apparatus becomes inactivated once they mature. It's hard to escape these spiders, as they're found in all forty-eight of the continental United States, in homes, garages, barns, and woodpiles.

You may not notice anything more than a pinprick at the moment a black widow bites you. Soon, however, you'll feel excruciating pain at the puncture site, severe abdominal pain, muscular pain in your legs, nausea, profuse sweating, headache, difficulty breathing, and swollen eyelids. After two to three days of being convinced you're on your deathbed, you'll recover—unless you're one of those rare persons who dies from respiratory paralysis.

The brown recluse spider, about the same size as a black widow, is the only other spider in the United States with venom dangerous to people. Because this spider hangs out in houses, people often encounter it—and get bitten. Like black widows, these spiders are not aggressive, but they do crawl into clothing. A bite produces a stinging sensation, and a few hours later the skin around the bite reddens and swells. Eventually the tissues die, and the open ulcer may take months to heal. Some people, especially children, have severe reactions to the venom. Most recover.

Many people are deathly afraid of tarantulas. Perhaps it's their size: some are the size of your palm, and one South American species has a leg span the size of a large dinner plate. Perhaps it's their hairiness. Or perhaps it's their fangs and venom, though they use them mainly for subduing prey.

Tarantulas stress easily, though, and they will bite to defend themselves. The first time I saw a tarantula give its defensive display, in the southern beech forest of Chile, I knew it meant business. I was sitting on the ground watching a Darwin's frog, when I moved my hand and brushed against something soft. I looked down and saw a fuzzy brown tarantula nearly the size of my hand. Frightened, it reared back, lifted its front two pairs of legs, and displayed its wicked fangs. I couldn't move or I'd frighten the frog I was watching. I stayed motionless and eventually the tarantula ambled off.

Although there are no documented human deaths from tarantula bites, venom injections from some tropical species can be painful. All tarantulas in the United States are docile and timid, and their bites are no more dangerous than bee stings. Many tame easily and can be safely handled. Frightened tarantulas, however, have a defense worse than their bite. They rub their rear legs up onto their abdomens and fling thousands of urticating setae (irritating hairs) toward the assailant. If you inhale these hairs, they attach to your mucus membranes. Some might penetrate your skin or get into your eyes. Some people experience intense discomfort and irritation from the setae, others only a mild itching. A small predator of tarantulas, though, might be blinded—or suffocate if it inhales setae into its lungs.

CENTIPEDES AND SPIDERS engender fear in humans because some have dangerous bites. Remember, though, they use their venom primarily to subdue active prey. Just as with venomous snakes, it's only when we bother these animals that they bite in self-defense. And they'd really rather escape than bite, despite what Hollywood would have you believe. They are beautiful and unique creatures, just protecting themselves in an often dangerous environment.

VENOM: SPINES, STINGERS, AND STINGING CELLS

sed nullum usquam execrabilius quam radius super caudam eminens trygonis quam nostri pastinacam appellant, quincunciali magnitudine; arbores infixus radici necat, arma ut telum perforat vi ferri et veneni malo.

[But there is nothing in the world more execrable than the sting projecting above the tail of the sting-ray which our people call the parsnip-fish; it is five inches long, and kills trees when driven into the root, and penetrates armour like a missile, with the force of steel and with deadly poison.]

PLINY, *Historia Naturalis*, vol. 3

Stingrays, often called "demons of the sea" or "devilfishes," have been known to be venomous at least since Pliny wrote his *Historia Naturalis* in the first century A.D., although I doubt the stinger could really kill a tree. These shark relatives range in size from pygmies only several inches in diameter to giants over fourteen feet long. They defend themselves by injecting venom through one or two sharp spines protruding from the upper side of the tail. The spine of a large stingray may be a foot long. Stingray spines are edged with hooks, which tear tissue as the ray yanks its spine back out. Stingrays use their spines only as weapons against sharks and other predators, not to capture their food (small fish, shrimps, squid, and other invertebrates).

Be cautious when strolling in shallow seawater. Stingrays lie buried in the sand. If you step on one, it can pack a wallop. A ray can swing its tail upward and jab its spine into you, resulting in a painful and serious wound that seems to take forever to heal even if it doesn't become infected—which it often does. The wound may be so severely lacerated that it requires stitches, but fortunately human deaths following stingray stings are rare.

One of my friends and her teenage son had an encounter with a stingray off the coast of a small fishing village in northern Ecuador. Carol was reading on the beach when suddenly her son limped toward her. He soon passed out, but not before gasping that he'd been hit by a stingray. No local doctor was available (they'd all flown to Quito for a physicians' holiday), so Carol took her son to the only medical person left in the village— a veterinarian. For the next hour they alternated soaking his foot in hot water, then cold water. The pain was almost unbearable at first but eventually subsided. To this day Carol's son tells the tale of the day his mother took him to a vet.

Spiny eels, pricklefishes, and sticklebacks can inflict painful wounds just by jabbing their spines into victims. Other bony fishes, such as certain catfishes, weeverfishes, stargazers, scorpionfishes, zebrafishes, and toadfishes inject venom through their spines.

Stonefish, grotesque twelve- to fifteen-inch creatures living in shallow ocean water of the Old World, are the most venomous of all fish. These broad-headed, warty fish have hypodermic-like spines on their fins: thirteen along the back, three in the anal fins, and one in each pelvic fin. Each spine is fueled by two venom sacs. The fish rest with their spines erect, camouflaged on coral reefs, where they mimic stones and wait to ambush small fish for dinner. If anything brushes against them, they inject a deadly nerve toxin.

An unwary person who accidentally steps on a stonefish gets jabbed and injected. He or she may have difficulty breathing, go into convulsions, and become delirious within fifteen minutes. Agonizing pain can last for twelve hours. If the fingers or toes are stabbed, these digits may turn black and fall off. If the leg is stabbed, it may swell to elephantine proportions and be unusable for three weeks or longer. Death may occur.

SOME ANIMALS—for example, scorpions—inject venom with stingers. Scorpions have curved stingers at their tail tips, and they usually hold their tails arched over their backs and poised for action. Two glands at the base of the stinger provide the venom that makes a sting painful, though rarely lethal for humans. There are exceptions, however. Every continent has scorpions potentially fatal to humans, including two species in the southwest United States. In the United States and Mexico, more people are killed every year by scorpions than by venomous snakes.

European honeybees defend their colonies by stinging intruders. The stinger is a modified ovipositor (egg-laying structure), which explains why only females can sting. When a honeybee jabs her barb-covered stinger into your toe or other vulnerable body part, the barbs cling to your flesh. As the bee jerks out her body to fly away, the stinger rips out, and it and the attached venom sac remain inside your toe. Your task is to remove the stinger as quickly as possible, since the venom sac can continue to pump out venom for another two minutes. There's a right way and a wrong way to do this. Scrape the stinger out with your fingernail. Don't pull it out—you'll just force more venom into your throbbing toe.

Unfortunately for the honeybee, her defense is a one-shot "use it and lose it" deal. After stinging an intruder, a bee is wounded so badly that she dies soon after flying away. Maladaptive, you ask? No. By leaving the sting apparatus in the flesh, all the venom is injected without the victim being able to remove the source (unless of course the animal is a human). Only the sterile worker bees attack and sting intruders, and their loss is no serious problem for the bee colony, which, like an ant colony, is analogous to one giant individual.

Africanized honeybees, closely related to European honeybees, have the same type of venom. They're no more toxic than their European relatives—just more aggressive. They're more likely than European honeybees to sting you because they're more unpredictable, faster to respond, and more likely to defend their nest. They also defend a larger area around their nest. You're likely to get more stings from the Africanized variety. If you disturb their nest, thousands of irate bees may pursue you.

Female wasps such as hornets and yellow jackets have barbless stingers, so they don't die after stinging you. This means they can sting repeatedly, though, so brush them off your skin and tiptoe away. Wasp venom contains an alarm pheromone (chemical signal), and when other guard wasps perceive the pheromone, they become agitated and aggressive. That's why if one wasp stings you, others may follow suit. Don't crush a wasp, as this will diffuse the pheromone even more and incite the guards to attack. Yellow jackets, like many bees, advertise their venomous nature. Analogous to yellow-and-black road signs that warn "proceed with caution," the yellow-and-black stripes of a yellow jacket warn: "Watch out! I have a stinger and if you cross me, I'll use it."

Bee and wasp venom is a complex substance consisting of histamines, proteins, and enzymes, among other components. When stung by a bee or wasp, most people experience temporary pain, swelling, redness, and itching. Not traumatic, just uncomfortable. An average nonallergic person would need 8.6 bee stings per pound of body weight before the venom became life-threatening: 1,290 stings for a 150-pound person. About one in every hundred people in the United States, however, is severely allergic to bee or wasp venom. Each time an allergic person is stung, the reaction is stronger, until eventually a sting may prove fatal. People with this allergy have a systemic reaction: dizziness, nausea, hives, difficulty breathing, swelling in the throat, drop in blood pressure, and anaphylactic shock. Between 30 to 120 people die each year in the United States from bee or wasp stings.

Whether or not you're hypersensitive to bee or wasp venom, if you don't want to attract these insects when you're hiking, gardening, or picnicking, don't look or smell like a flower. Don't wear brightly colored clothing, and forgo the perfume, hair spray, and after-shave lotion.

About half of the approximately ninety-five hundred named species of ants have rear ends armed with stingers that inject venom into prey or assailants. Argentine fire ants, Australian bulldog ants, and others have stingers strong enough to pierce the skin of humans and other mammals. Some species, though, have reduced stingers effective only against other insects.

For humans, one of the most crippling ant stings is from the giant neotropical conga ant (*Paraponera clavata*). David Pearson and Les Beletsky write in their ecotravelers' wildlife guide to Ecuador: "The large . . . all-black conga ant . . . is probably the most dangerous animal you will encounter during your trip to lowland rainforests in Ecuador. If all you do is see it, count yourself lucky, as the sting of this animal is considered one of the most painful in the entire world—it may even cause hallucinations in some people." These large one-inch-long ants often forage singly along the ground and in vegetation. They hunt both live prey, such as other insects and sometimes small lizards, but they also scavenge dead animals and eat droplets of nectar. I can attest to their painful sting.

While doing fieldwork in the rain forest of Ecuador, I was stung several times by conga ants. Once I put on my flip-flops to answer nature's call in the middle of the night and got nailed on the toe. Sharp bursts of pain shooting through my foot and up my leg kept me awake the rest of the night. Another time, I accidentally brushed my hand against a conga ant and got stung in the finger. My arm swelled to twice normal size; tears flowed uncontrollably from my eyes. I couldn't use my hand for nearly two days. Locals told me I was lucky. Many people are laid up with burning fever for forty-eight hours following a conga ant sting, and some have life-threatening reactions to the venom.

AND THEN THERE are stinging cells. Many members of the phylum Cnidaria (from the Greek word meaning "nettle")—including sea anemones, jellyfish, stinging corals, and fire corals—have stinging cells that contain minute harpoonlike capsules called nematocysts that explode when touched. If you brush against some of these cnidarians, the experience may remind you of tangling with a nettle, but others give a more powerful punch. The sea wasp (a type of jellyfish) shoots venom deadlier than any snake venom from its nematocysts—a venom that can kill a person within three minutes.

Another nasty jellyfish is *Cyanea capillata,* the villain of the Sherlock Holmes story "The Adventure of the Lion's Mane." Sir Arthur Conan Doyle describes the jellyfish as "a tangled mass torn from the mane of a lion. It lay upon a rocky shelf some three feet under the water, a curious waving, vibrating, hairy creature with streaks of silver among its yellow tresses. It pulsated with a slow, heavy dilation and contraction." One person in the story gets stung by the jellyfish but doesn't die. Doyle writes that he was "dazed in mind, and every now and then he explained that he had no notion what had occurred to him, save that terrific pangs had suddenly shot through him. . . ."

Perhaps the most spectacular cnidarian is the Portuguese man-of-war, named by English sailors who saw these creatures in the seas of Portugal and thought they looked like the sailing ship called man-of-war. The beautiful but dangerous violet-blue Portuguese man-of-war, found in warm tropical seas, resembles a jellyfish but is actually composed of hundreds of individual organisms. Each organism is specialized for a specific function such as digestion, defense, or reproduction. Part of this super-creature drifts on the water surface: a twelve-inch gas-filled structure called a float. Attached to the violet-blue balloonlike float and trailing into the water below are tentacles up to thirty feet long. The float acts as a sail and catches air currents, allowing the animal to move through the water entirely by the wind and waves. Like its relatives the sea anemones, the Portuguese man-of war discharges nematocysts on its tentacles and fires venom into a fish or other prey, paralyzing or killing it. The tentacles then draw the prey up into mouths of gastrozooids (feeding individuals), where they digest food for everyone in the colony. Food sharing is accomplished thanks to intercommunicating chambers that serve for circulation, digestion, and distribution of food (gastrovascular cavities).

A Portuguese man-of-war also fends off predators with its stinging cells. If one of these super-creatures brushes up against your skin, you'll experience pain, welts and severe rash, nausea, and difficulty breathing. These animals are often washed up onto the beach. Beware: even dead they remain dangerous to touch.

SO HOW TO deal with venomous stings? Let's return to Pliny, the first-century Roman scholar, and his thirty-seven-volume *Historia Naturalis*. Perhaps if one strongly believes in Pliny's suggestion that a broth of boiled tortoise flesh will relieve the pain of a weeverfish sting, it will. Ancient Romans and people during the Middle Ages relied heavily on Pliny's recommendations. As remedies for scorpion sting, Pliny suggested several options:

- Boil a she-goat's dung in vinegar and place over the wound.
- Drink a mixture of beaver oil and wine.
- Apply river snails preserved in salt.
- Pour a man's urine over the sting site.
- Cover the wound with pounded earthworms.

WE TEND TO view stinger venom as all bad, but it has a good side also. Just as snake venom can be used for medicinal purposes, so can bee venom. Apitherapy, or the use of honeybee products for therapeutic purposes, goes back several thousand years to ancient Babylonia, Egypt, Greece, and China. It's been around ever since. Hippocrates, the Greek physician often called the father of medicine, treated arthritis-related diseases and other painful joint conditions with bee venom. During the Middle Ages, Charlemagne and many others reportedly used bee venom to treat their aching joints. During the sixteenth century, Ivan the Terrible of Russia cured his gout by allowing bees to sting his affected areas. Bee venom therapy continues into the twenty-first century, especially in Asia, eastern Europe, and the former Soviet Union, where it's widely practiced. In case you're tempted, be advised that not all patients obtain relief from their symptoms when injected with bee venom—yet all feel the pain.

What's the basis for honeybee venom's purported medicinal value in treating inflammatory diseases? Basically, it's thought to work by stimulating the body's immune system. A bee's sting causes inflammation, which triggers your adrenal glands to release cortisol, a natural hormone that reduces inflammation. By allowing a bee to sting an area that is already inflamed, the cortisol released to deal with the bee venom is so strong that it could treat the arthritis, tendonitis, bursitis, or other inflammatory condition at the same time.

Bee venom therapy can be administered either through a hypodermic needle or directly through a bee. In the latter case, a bee is held with tweezers over the affected area and allowed to sting. To lesson the pain, the area to be treated can be iced both before and after the stings. The quantity of venom needed depends on the problem. A simple tendonitis might require only a few stings per session for up to five sessions. Arthritis might require several stings two or three times per week for up to several months.

AND SO YOU see that venom from a bee stinger isn't all bad. Chances are good that you don't react strongly to it (unless you're one of those one in a hundred), and it just might have medicinal value—at least for some people, some of the time.

But watch out for those stingrays. Pliny recommended applying a compress of ray ash soaked in vinegar onto the wound. If your local pharmacy doesn't carry that, you could always ask a veterinarian what he or she recommends.

BORROWED AND STOLEN DEFENSES

Her little black nose went sniffle, sniffle, snuffle, and her eyes went twinkle, twinkle; and underneath her cap—where Lucie had yellow curls—that little person had PRICKLES!

BEATRIX POTTER, *The Tale of Mrs. Tiggy-Winkle*

Something about hedgehogs has intrigued me ever since my mother first read Mrs. Tiggy-Winkle's story to me. Twenty years later, I learned firsthand about hedgehogs when I lived in Brooklyn for a time and a biologist friend gave me a baby hedgehog for a pet. I named the hedgehog Monsta.

Early on, Monsta showed me how a hedgehog uses its "do not touch" prickles. He was exploring my apartment's expanse of cream-colored shag carpet when my boisterous neighbors began shouting obscenities and throwing pots and pans. Monsta tucked his vulnerable snout and feet up against his belly, erected his spines, and rolled into a bristling pincushion. He hissed and shook. I bent down to comfort him, and the vibrating ball of spines lurched and hit my bare foot. I yelped and jumped backward. His defense had worked.

Over the next few days, Monsta also showed me his capricious nature. Capricious: apt to change suddenly or unpredictably; to change one's mind impulsively; from the Italian *capriccio,* meaning "head with hair standing on end"; from *capo,* meaning head (Latin *caput*), and *riccio,* meaning hedgehog (Latin *ericius*). Monsta was definitely capricious. One minute he was a relaxed, snuffling animal trundling about in search of food. The next he was an antisocial hissing ball of spines. He was also sometimes irritable and unapproachable, leading credence to the expression "prickly as a hedgehog." I was fascinated by Monsta's behavior.

One evening after a long bout of housecleaning, I carefully lifted Monsta from his aquarium and set him on the carpet for his daily exploration. No doubt he smelled something unusual, for he licked my hand. Monsta foamed at the mouth, then raised up on his stubby front legs, craned his head over his shoulder, and smeared the white froth onto his spines with his tongue. After a minute he turned and frothed his other side. Licking a foreign substance and then anointing itself with saliva is a hedgehog's natural response to novel, smelly, or noxious substances—in this case a residue of cleaning fluid on my hands. Why do they do this? Smelly or toxic substances smeared onto the spines might render the spines even more effective weapons against enemies.

Before becoming my pet, Monsta was a laboratory "guinea pig" who played a key role in experiments with toads. Because toads live in the same environments as hedgehogs, my biologist friend Butch Brodie wondered if hedgehogs might anoint themselves with toad poison. He put some toads and hedgehogs together and got his answer. The hedgehogs bit the toads and chewed the large poison-filled glands on the toads' heads. After foaming at the mouth, the hedgehogs spread the combination of saliva and toad poison over their spines. Many of the hedgehogs then ate the toads, unaffected by the toxins that kill most would-be toad predators.

Butch wondered if spines anointed with toad poison would be more painful than non-anointed spines when they penetrated a predator's skin. How better to determine this than on oneself? But a scientist needs a reasonable sample size, so Butch enlisted the arms of six "volunteer" graduate students. (This experiment took place in the mid-1970s, before the proliferation of personal-injury lawsuits.) Butch removed some spines from a hedgehog and treated them in three ways: (1) washed in alcohol; (2) washed in alcohol, then coated with hedgehog saliva; and (3) washed in alcohol then coated with toad poison. Butch removed a fourth group of spines from the hedgehog after it had anointed itself with toad poison. Each of the six students and Butch jabbed him-/herself in the inner forearm with one of each of the four types of spines in random order. Only Butch knew which spine was which.

The results were compelling. In only one of the seven subjects did spines of the first two treatments cause any reddening, and there was no pain or burning. In contrast, six of seven subjects poked with spines from the third treatment group experienced immediate intense local burning and splotchy red areas around the puncture sites. All seven persons experienced these same effects when jabbed with spines removed from the hedgehog that had anointed itself with toad poison. The burning sensation from the third and fourth treatments lasted up to sixty minutes. Experiments carried out a week later with dried spines that the hedgehog had self-anointed still had the same intense effect.

As critical as spines are for defense, the sparse whitish spines of newborn hedgehogs are too soft to be functional—no doubt a well-devised plan by Mother Nature, who sided with mothers on the birthing details. A few days after birth, a second coat of darker spines pops up between the flexible ones. At about two weeks of age, the babies open their eyes, and a third set of rigid, sharp dark spines grows through the skin. At about one month, the young shed the first two sets of spines. Soon the babies are out and about, following their mother as she snorts and snuffles beneath hedges searching for insects, earthworms, slugs, and snails to eat. At this point, the youngsters can roll up into tight little balls to protect themselves, and they can even foam at the mouth and anoint their spines.

HEDGEHOGS AREN'T ALONE in borrowing from other animals for their own defense, and toads aren't the only lenders. The natural defenses of some members of the phylum Cnidaria are highly sought after. Cnidarians (jellyfish, corals, hydroids, sea anemones, and their relatives) have stinging cells, which other animals exploit.

One peculiar animal whose defense is highly sought after is the sea anemone. These marine invertebrates have tentacles bearing stinging cells. Each stinging cell contains a minute harpoonlike capsule called a nematocyst. When a predator grabs onto a tentacle, the nematocysts discharge and the tiny daggers inject a powerful nerve poison.

Clownfish are cheery-looking bright orange fish with three white bands around the body. Their outlandish appearance is more than matched by their fascinating behavior: they hide out among the tentacles of sea anemones for protection. Although the anemones' stinging cells are effective defenses against other fishes, clownfish are covered with a special mucus that protects them from the nematocysts. When a prey unexpectedly touches the sea anemone's tentacles, the nematocysts discharge and the anemone has caught itself a meal. Grateful for a free handout, the resident clownfish feeds on scraps left behind by the anemone. The clownfish leaves the anemone—briefly and not too far afield—to gather its own food, such as filamentous algae, but returns to the safety of the tentacles to eat. Conveniently, the sea anemone consumes bits of food that drift down from the fish's mouth. The two share a mutually beneficial relationship.

The Walt Disney animated movie *Finding Nemo* illustrates just how dangerous life out of the anemone can be for clownfish. The dad, Marlin, and his sidekick Dory, a forgetful regal blue tang, encounter sharks, a whale, an anglerfish with a bioluminescent lure, and a forest of jellyfish as they travel through open water in search of Marlin's son, Nemo. (Nemo has been netted by a scuba diver and ends up in a saltwater tank in a dentist's office.) Despite the fact that a father clownfish would never search for his son, the dramatic and colorful scenes of corals, sponges, crabs, fish, and sea turtles on Australia's Great Barrier Reef will no doubt inspire a new generation of marine biologists. As for how Nemo escapes the dentist's office . . . you'll have to watch the movie to find out.

Hermit crabs wriggle and wedge themselves into empty snail shells to protect their soft abdomens. A crab grasps on to its borrowed shell with a pair of modified legs. If you've ever tried to extract a hermit crab from a beautiful shell *you* wanted, you know these little crabs are tenacious. Hermit crabs often camouflage their shell homes. One clever disguise involves adopting a sea anemone. A hermit crab nudges a sea anemone's body and coaxes it loose from a rock. The crab then carefully lifts the anemone and places it onto its shell. A large hermit crab often decorates its shell with more than one anemone. Now, instead of thinking the creature is an edible crab, predators will assume it's just a venomous anemone. In addition to providing a passive

disguise, the anemone provides active defense for the crab. If an octopus, squid, or other predator gets too close, the anemone's stinging cells will discharge. In return, the anemone gets a free ride, probably allowing it to eat more than if it were still attached to the immobile rock.

My favorite animal that uses anemones for its own protection is the boxer crab. These resourceful crabs hold a sea anemone in each claw as a shield as they cruise through their environment. With their anemone-fitted claws held up, the crabs resemble boxers primed for a fight. When a predator approaches, the crab lunges and threatens with stinging tentacles. If the enemy doesn't retreat, it gets punched with discharging nematocysts.

Hydroids (small colonial plantlike animals) also have nematocysts. Despite this, some sea slugs (nudibranchs) can eat hydroids and sea anemones without causing the nematocysts to discharge. The stinging cells pass unharmed through the sea slug's digestive system and eventually collect in specialized sacs. These sacs open to the outside by pores at the tips of the numerous outgrowths covering a sea slug's body. Although the sea slug can't discharge its sequestered nematocysts at will, the cells discharge when something touches the animal. Thus, the nematocysts that once protected hydroids and sea anemones now protect sea slugs from their own enemies.

Another potent nematocyst-laden cnidarian is the Portuguese man-of-war. It's hard to believe, but the small blue-and-silver-striped man-of-war fish lives within the deadly tentacles of this large super-creature. Like the clownfish gaining protection from its sea anemone, the man-of-war fish uses the Portuguese man-of-war as a home, a safe haven from the dangers of the open sea.

NOW YOU KNOW what toads and cnidarians have in common: antipredator defenses so effective that other animals borrow or steal them. Beatrix Potter never revealed to us that Mrs. Tiggy-Winkle was the sort to sink her teeth into a toad, foam at the mouth, and spread poisonous saliva over her spines. Then again, if this charming early twentieth-century author had known about hedgehog saliva tinctured with poison obtained by chewing toads' heads, she probably would have stuck to writing about bunnies and kittens. It took Monsta and his friends to show us the truly enterprising nature of hedgehogs.

MASQUERADING AS DEBRIS

The Wolf in Sheep's Clothing

In order to make easy work of hunting for his supper, a clever wolf struck upon the idea of dressing himself as a sheep. When hungry, he put on a sheepskin and mingled with the flock. So good was the disguise that the shepherd did not notice anything amiss, and the wolf could choose his prey at his pleasure. One evening, however, the shepherd desired mutton stew for his supper, chose a suitable sheep from the flock, and slaughtered the animal—only to discover that his intended dinner was a wolf in sheep's clothing.

The Fables of Aesop, retold by F. BARNES MURPHY

What fun to hide behind a mask, don a costume, and disguise ourselves for Halloween or a masquerade party. For an evening we can pretend to be someone or something we're not: ghost, witch, Darth Vader, Elvis, bumblebee, elephant, hot dog, rainbow, cadaver, whatever.

Although humans are the only animals (outside of Gary Larson's cartoons) that disguise themselves for pure frivolity, some nonhuman animals deliberately decorate their bodies with foreign material for a serious reason. This behavior, called masking, has at least three possible functions. The first two fit into this chapter as ways animals protect themselves from being eaten. First, masking may allow the animal to blend into its environment and go unnoticed by predators. Second, it might transform the animal into what appears to be an unpromising food item. The third function of masking, and the one illustrated by Aesop's fable, belongs in the feeding chapter, as the disguise might allow the animal to sneak up on its prey. Because the same disguise may function in two or even all three ways, though, we'll look at all the masking behaviors in this chapter.

ONE EXTRAORDINARY EXAMPLE of masking involves green lacewing insects. Adult lacewings are delicate creatures with two pairs of lacy wings. The larvae, the stage that mask, have hairy elongated bodies tapered at both ends; their heads are armed with strong sickle-shaped jaws. Thomas Eisner and his colleagues reported a "wolf-in-sheep's-clothing" behavior in 1978. The sheep in this interaction are aphids, and the wolves are larval lacewings.

Aphids (also called plant lice) are tiny soft-bodied insects despised by gardeners and farmers because they pierce plant tissues and feed on fluids from stems, leaves, flowers, and even roots, causing the plants to wilt. Aphids consume huge quantities of fluids, excreting the excess carbohydrate as a sweet liquid called honeydew. This sugary liquid doesn't go to waste. Many insects, especially ants, love it. In fact, ants have "struck a deal" with aphids. In return for protecting aphids from predators and shepherding them from one plant to another, ants are allowed to drink honeydew at the source: they lick up droplets from the aphids' rear ends. Even more bizarre, ants often solicit honeydew by stroking aphids with their antennae.

So the dilemma for an aphid predator is how to snatch an aphid without getting bitten by ferocious bodyguard ants. Larval green lacewings (the wolves) eat aphids (the sheep), and they've solved this dilemma in a clever way: they masquerade as aphids.

Green lacewing larvae live in colonies of wooly alder aphids. These aphids are so-named because they suck fluids from alder bushes and because they have a fluffy sheeplike appearance. The fluffy appearance comes from tufts of thin white filaments of wax the aphids secrete onto their bod-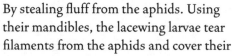ies, making themselves conspicuous against the dark alder branches—all the better for the ants to see them. How can a naked lacewing larva blend in? By stealing fluff from the aphids. Using their mandibles, the lacewing larvae tear filaments from the aphids and cover their own bodies with this waxy material by pressing it down with their heads. Long bristles on the lacewings' backs and sides secure the wax in place. Curiously, the aphids offer no resistance to being defluffed.

Lacewing larvae are about the same size as the aphids, and with their cover of white fluff, you'd swear they were wooly alder aphids. If the wax shields are removed from the lacewing larvae, the ant bodyguards attack and grasp the naked intruders in their mandibles. The ants pull until the larvae lose their footing and then drop them to the ground. Of course, under normal circumstances, no one denudes lacewing larvae. Disguised in their sheepskin, these wolves can roam undetected among the flock of sheep, picking them off one by one for supper.

THE NYMPH (JUVENILE) stages of some assassin bugs are also clever masqueraders. Nymphs of several West African assassin bugs disguise themselves in a two-part costume. To begin with, a bug covers itself with sand, soil, and dust particles, which stick to a dense layer of bristles covering its body. Dust and dirt conceal the bug's natural body odor, allowing it to lie in wait for its prey, mainly ants, without being detected by smell.

The second part of the costume protects the nymphs from predators. Assassin bugs have earned their common name because they pierce their beaks into prey and inject saliva that paralyzes their victims almost instantly. Digestive enzymes in the bugs' saliva dissolve the prey's tissues. The bugs suck out the liquids and leave behind hollowed-out shells. But these particular West African assassin bugs don't always discard the corpses. As the second costume layer, a nymph piles prey corpses and other bits of animal and plant matter onto its back. The "backpack" is held in place by fine elastic threads secreted by glands on the bug's back. After a bug molts, it often takes up its old backpack. Sometimes it even grabs its shed skin and adds that to the debris it carries around.

The behavioral ecologist David Owen observes that "nothing looks less like a bug than a small heap of ants." Assassin bugs running around with empty shells of ants, termites, and other insects no doubt confuse spiders, centipedes, geckos, and other predators. Ideally, the predator won't recognize the roving heap of debris as food. If the bug is attacked, however, it can shed its backpack—leaving a predator to inspect the empty husks while the bug escapes. The backpack plays the same role as the self-amputated tail of a salamander or lizard: distraction.

Another kind of assassin bug lives on carton nests of Costa Rican termites. Carton is wood that worker termites have chewed and then cemented back together with their feces. Nymphs of this particular assassin bug scrape carton from the termites' nests and then pat the crumbs onto their bodies, which are covered with tiny glue-secreting bristles. This costume camouflages the nymphs from their visually hunting predators: they look like fragments of chewed-up wood, not assassin bug nymphs.

But the costume does more than protect the bugs from their enemies. The bugs not only look like bits of termite carton; they smell like carton—a handy disguise because the nymphs eat termites. By smelling like the nest, the bugs can approach their food without getting attacked by blind soldier termites that protect the nest from intruders.

The deception goes further. Worker termites generally hang out inside the nest, repairing damage and enlarging the structure. The bugs entice the termites to leave home by baiting them. Worker termites are strongly attracted to dead and dying nest mates. When they detect them, the workers exit the nest and eat their former companions—a nutrient-conserving strategy if you will. The bugs take advantage of this behavior by dangling termite carcasses where the nest needs repair. When a worker termite leans out of the crack or hole to grab the bait, the bug drags the squirming termite out and sucks it dry.

MANY SPIDER CRABS, so-called because they somewhat resemble spiders with their pear-shaped bodies, camouflage themselves from predators with seaweed, bits of animals, or other debris. Methodically, a crab picks up foreign material in its claws, mouths and softens the debris, and then rubs the decor onto its body where hooked bristles hold it in place. Some spider crabs pile additional cover onto themselves when they move into a different habitat, allowing them to fade into the new surroundings. After a meal, some species pick up their leftovers—seaweed and other food—and attach the morsels to their bodies. The leftovers provide camouflage, until the crabs later eat the dregs of their meals.

Some crabs play dress-up when young but then outgrow the game as adults. Juvenile kelp crabs (a type of spider crab) actively pile marine algae onto their bodies, which are covered with hooked bristles. Once they mature and are large enough to defend themselves, they lose the bristles and the ability to decorate.

Spider crabs make fascinating aquarium pets, but beware. They're inclined to tear up their tank mates (the anemones and sponges you've paid a fortune for) and attach them to their bodies.

THE NEXT TIME you wrap yourself in a costume to disguise your identity, remember the green lacewing larvae that steal white fluff to camouflage themselves; the assassin bugs that carry backpacks of debris; and the spider crabs that cover themselves with colorful seaweed. Better yet, use their disguises for unique Halloween costumes or for that masquerade party. Cover yourself with cotton balls, croutons, or seaweed. You might just win the best-costume prize!

5 Ya Don't Say!

ON WET SPRING and summer evenings, the raucous croaks, peeps, ribbits, trills, bonks, squawks, and whistles emanating from ponds may be deafening to human ears, but they are music to female frogs. Each male sings: "Come to me, I'm the best." Other animals have their own ways of saying the same thing. A male peacock woos his lady love by parading and spreading his metallic green train of feathers into a magnificent fan. Snails communicate their romantic intentions with tentacle caresses and erotic stabs. Boars produce a musky-smelling steroid in their testes; this chemical is released into their blood and reaches their salivary glands. During courtship, males foam at the mouth. Females, enraptured by the smelly drool, arch their backs and spread their legs, effectively saying, "I'm yours."

Acoustic signals, visual displays, touch (tactile signals), and odors (chemical communication) are the four main ways that animals attract mates. Beyond the communication necessary for mating, animals also use these means to relay information to advertise ownership of territory, warn of danger, share information about food, and maintain group contact. In this chapter, we'll look at some unusual forms of acoustic, visual, tactile, and chemical communication.

Acoustic signals: songs, calls, and other vocalizations and sounds allow communication over long distances, such as within prairie dog towns. Prairie dogs, the most social of all rodents, have an especially sophisticated language.

Visual signals: reflecting our human bias, we usually think of visual signals as working best during daylight hours. Some nocturnal animals and some that live in the oceans, however, use lights to attract mates; startle, confuse, or frighten predators; or lure food.

Tactile signals: this form of communication functions well at short range. We don't usually think of reptiles as being physically "affectionate" with each other, but some are. Others use tactile signals in a more aggressive way: they bite their lovers. Either way, reptiles communicate complex messages through touch.

Chemical signals: pheromones, another powerful form of communication, convey many types of information, from alarm messages to road maps leading to food sources to sexual lures.

People communicate differently to strangers than to friends. Some other animals do also. We'll end the chapter by looking at several animals that respond less aggressively to neighbors that cross into their territories than to intruding strangers. They recognize neighbors by their song, the smell of their feces or glandular secretions, or their appearance.

TAKE COVER!

A coyote is hunting. He quickly approaches the edge of a colony of prairie dogs hoping to catch an unwary animal. But the sharp-eyed prairie dogs spot him, and a chorus of barks rises into the still morning air. Some prairie dogs who have been breakfasting on grass dash to the mouths of their burrows where they stand on their hind legs watching the marauder. Others who have been resting inside their burrows emerge, and they too stand looking at him. The coyote walks slowly. He is alert for a prairie dog who might be just a little too far from its burrow, one he can charge, perhaps confusing it into running in the wrong direction. But all the inhabitants of this colony are standing a single bound from their burrows with their eyes fixed on him, and so he decides to leave, still hungry, in search of easier prey. The prairie dogs continue to stand at attention, watching him go, barking at him as he disappears into the grass at the edge of the colony.

CON SLOBODCHIKOFF, "The Language of Prairie Dogs"

Prairie dogs, members of the squirrel family, are the most social of all rodents. Sometimes more than two thousand individuals live in a colony, or town. A prairie dog town encompasses many coteries, or family groups, each of which includes an adult male,

his harem of up to six adult females, and their offspring. Members of a family group spend most of their time within a territory of usually less than one acre, which includes up to a hundred burrows, some fifteen feet deep. Most burrow complexes have at least two entrances, ensuring that the residents have an escape route if a predator, such as a rattlesnake, enters the burrow. These rodents are great little communicators, and they "talk" with each other in many different contexts.

Prairie dogs greet each other by touching lips, with mouths partially open and teeth exposed. They often touch teeth, sometimes even locking their incisors together, for up to ten seconds. This smooch is one way of identifying each other. After the kiss, animals belonging to the same coterie wag their tails and groom each other or play for a while. If the animals are strangers, they raise their tails, expose their anal glands, and take turns sniffing each other's rear end. Usually the animal that doesn't belong in the territory flees. If it doesn't, the two may fight—sometimes a bloody battle, though rarely fatal.

A prairie dog announces its territory to the world by standing on hind legs and belting out a dramatic two-part doglike bark as it breathes in and then exhales. Sometimes an individual barks so forcefully that it leaps into the air, flips, and topples over backward. If an intruding prairie dog approaches, the resident gives a slow intermittent bark and often chatters its teeth. As the interloper flees, it typically utters a churring sound, perhaps signaling its submission.

Boundaries of coteries are passed on through generations. Each immature prairie dog learns the territory boundaries through its actions with other individuals. When it's inside the boundary, it participates in mutual grooming with other coterie members, but when it strays outside the boundaries, it encounters aggression and rejection through squabbles and skirmishes.

As indicated in the quote, prairie dogs communicate with each other in another context: they warn of impending danger. Prairie dogs suffer from frequent predation. An individual has only a fifty-fifty chance of celebrating its first birthday. Indeed, prairie dogs spend up to 43 percent of their awake time scanning for predators—eagles, hawks, badgers, coyotes, domestic dogs, and humans. When a predator approaches a prairie dog town, all hell breaks loose. The first prairie dog alerted to the danger yips a high-pitched warning that soon spreads like a wave from burrow to burrow and may send the residents bolting for cover.

Con Slobodchikoff and his students at Northern Arizona University have discovered that Gunnison's prairie dogs (the species found on the Colorado Plateau) can distinguish between different predators. They give distinct alarm calls depending on the particular predator and risk level. The human ear can distinguish the different calls, but Con and his collaborators have also analyzed the calls from tape recordings they've made. The particular combination of call characteristics (for example, duration, frequency, rate of frequency modulation, and harmonic structure) allows the researchers to further distinguish among alarm calls.

A loud, sharp single-note alarm call means "Hawk overhead," and all the prairie dogs in the bird's flight pattern dive into their burrows. Nearby prairie dogs may stand at their burrow entrances and gawk, like humans at an accident scene. A longer and less sharp single note combined with bouts of calling (sets of numerous barks separated by a pause) from only one individual at a time means "Human approaching." In response to this call, the prairie dogs run to their burrows but head underground only if the person gets too close for comfort. Coyotes are better predators than domestic dogs and thus more of a threat. When several prairie dogs simultaneously give repeated bouts of calls, the message is "Dog alert!" The town's citizens freeze in place and stand up on back legs, assessing the threat. If the dog comes too close, they race to their burrows and disappear. In contrast, waves of bout calling from numerous individuals across town means "Coyote hunting," and the prairie dogs race to their burrow entrances, where they stand alert on their hind legs and scan for the enemy. If you play a tape recording of one of these four types of calls, prairie dogs exhibit the appropriate response even when there's no predator around. If you play a tape with nonsense sounds, the prairie dogs ignore it.

OTHER MAMMALS ALSO vary their alarm calls. Several kinds of monkeys, dwarf mongooses, ringtailed lemurs, and red squirrels utter different calls when they detect aerial versus terrestrial predators.

Vervet monkeys from the savannahs of Africa give distinctly different alarm calls for leopards, eagles, and snakes. Nearby vervets react to these various alarm calls differently. Monkeys on the ground run up into the trees when they hear leopard alarms. Individuals in the trees respond to eagle alarms by looking up into the sky, running down out of the trees, and seeking cover on the ground. Monkeys hearing snake alarms scan their surroundings. As with prairie dogs, if you play a tape of a specific alarm call to a vervet in the absence of a predator and a caller, it generally responds appropriately. Depending on the call, it runs up a tree, runs down a tree, or scans its surroundings.

BUT LET'S RETURN to prairie dogs. Gunnison's prairie dogs incorporate information into their alarm calls about the physical features of individual predators. Consider human predators. People hunt prairie dogs for food, shoot them for target practice, and poison them to eliminate their burrows and to prevent bubonic plague. When different people walk through a prairie dog town, the inhabitants include information in their alarm calls about the color of clothes and the size and shape of the intruding person. Con and his collaborators were amazed to find that the prairie dogs they studied gave the same call for a man wearing the same clothes after not having seen him for two months!

The prairie dogs also include information on the level of threat. In trials of one experiment, a man wearing a lab coat walked through a prairie dog town lunging at the rodents with a pretend gun—acting rather offensively. In different trials, the same person, again wearing a lab coat, walked through the colony with his hands in his pockets—projecting a nonthreatening image. The prairie dogs' calls were consistently different in response to these two human behaviors.

Wouldn't it be fascinating to break the prairie dog code? Can't you imagine one warning the others, "Run for cover, here comes the skinny kid in the blue T-shirt!" An hour later, one announces, "Here's that tall woman again." Later in the afternoon, a prairie dog barks out, "Watch out, here comes the fat wiener dog." And awhile later, one says, "The dopey sheepdog is back."

The advantage of distinguishing between individuals within a class of predator (for example, people or dogs) may be that individuals vary in their hunting behavior and success. Perhaps the skinny kid routinely aims his slingshot at the prairie dogs and is a greater threat than the tall woman who walks through town and just watches them. And perhaps the fat but spry dachshund is a greater threat to the residents than the shaggy sheepdog that bumbles through town.

Why should a prairie dog warn others of danger when the call might attract attention to itself and increase its own chance of getting eaten? Studies suggest that in general, individuals with relatives within earshot are more likely to give alarm calls than are individuals without kin nearby. Females are especially likely to give warnings when their youngsters are aboveground. It may take a village to raise a prairie dog, but each individual primarily looks after its own.

Gunnison's prairie dogs also seem to "chat" with each other. Con and his collaborators have found that prairie dogs around Flagstaff give distinct barks when they see certain large mammals, even though these animals don't eat prairie dogs. When a prairie dog barks the message "Cow," it seems to be just an informational item, as none of the town residents run or show any concern. The same is true for barks indicating approach of antelope, elk, or deer. Even more remarkable is that these prairie dogs coin new "words" for objects they've never seen. When the investigators wheeled a 32-by-26-inch black oval plywood shape across the colony on a pulley system, all the prairie dogs gave the same unique call—even though they don't routinely see big black ovals cruising through town.

What else might they talk about? The weather? Do they gossip? Clearly, we're just beginning to understand prairie dog language.

LIVING FLASHLIGHTS

Lightning Bug

A lightning bug is crazy.
It hasn't any mind.
It travels through existence
With its head light on behind.

GELLETT BURGESS, from *Nature Smiles in Verse*

The night air radiates twinkling yellow lights, dancing across the meadow. Close to the ground, other yellow lights flash in response. Fireflies act out this magical scene on warm summer evenings the world over.

Also called lightning bugs, fireflies are neither flies nor bugs. They're soft-bodied beetles. The light-producing organs responsible for the green, yellow, or orange flashes are usually located on the underside of their abdomens. The number, size, and arrangement of these organs varies from species to species.

Male fireflies fly around, flashing their signals to seduce females. Each species flashes in a distinctive pattern. The duration, time interval, and number of flashes identify the species so that only its own members recognize it as a mating signal. Females perch on the ground or in bushes and wait until a male flies nearby, flashing his offer for romance. If receptive, a female recognizes the signal as her species and answers with her own flash. Males recognize females of their species based on how long it takes a female to answer with a flash and the duration of that flash. The two continue their conversation of repeated flash-answer sequences until the male reaches the female.

In the world of fireflies, though, all is not always what it seems. On summer evenings in eastern North America, females of at least twelve species of the firefly genus *Photuris* mimic the flash responses of females belonging to other genera. Visualize the following. A male firefly, flitting about advertising for sexual favors, suddenly sees an answer broadcast from below. He zeros in on the spot and reinforces his message. The female repeats the correct answer a second time. He signals before alighting on the branch, and the female flashes her final answer. The romantic mood is shattered when the much larger female pounces on and devours her duped victim—a femme fatale in action. Why don't these aggressive females devour males of their own species? Timing. It's not until several days after they've mated that the *Photuris* females develop predatory tendencies. And because they've already mated, they no longer respond to the flashes of males of their own species.

We're all familiar with twinkling lights dancing across meadows at night, but how do fireflies produce their light? In 1885 a French physiologist, Raphael Dubois, removed the light-producing organs from some glowing beetles and clams, did some experimental manipulations, and demonstrated that, in addition to oxygen, at least two chemical compounds are required to produce the light. A chemical reaction takes place: an enzyme reacts with a photochemical and converts chemical energy stored in the cells into light without producing heat. These compounds were subsequently named luciferin (the photochemical) and luciferase (the enzyme), after Lucifer, the Latin word for "bearer of light." This chemical process—light given off by means other than heat—is called luminescence, and in living organisms it's called bioluminescence.

FIREFLIES AREN'T THE only bioluminescent animals that use lights to communicate with each other prior to mating or to lure prey, and certain other bioluminescent animals use their lights for still other reasons: to escape or avoid predators, or literally to light their way.

Although bioluminescence occurs in at least eight families of four orders of insects, it is most common among the beetles, including the spectacular headlight beetles found in the Caribbean and Central and South America. In some species, two round greenish-yellow spots on the side of

the head end glow steadily like headlights. Others also have a smaller tail-
light at the back end that glows orange or red, prompting a common name
of "Ford bug" or "Jeep." One Jamaican click beetle has paired green or
yellow-green headlights and also a yellow-green or orange light on its
underside. These beetles use their lights to attract mates, but they don't
flash them as fireflies do. Biologists have speculated that some of these
night-flying beetles may also use their glowing organs as "landing lights."
Local people have ingenious uses for headlight beetles: A dozen in a jar
provide enough light to read by; a few placed in a gauze bag clipped to hair
make a charming barrette; clear plastic bags filled with beetles and tied to
ankles function as superb living flashlights while walking through the forest at night.

Some soil-living animals produce light. A glowing light crawling along
the ground at night could be a bioluminescent centipede. Because most
glowing centipedes are blind or nearly so, they clearly aren't using lights to
attract mates. Instead, the light may serve as a warning and protect these
centipedes from their enemies. When attacked by a predator, the centipede
secretes a bioluminescent slime, which has a repulsive odor and burns on
contact. After a few unpleasant encounters with the caustic slime, a predator might learn to avoid crawling lights. Bioluminescent earthworms ooze
a glowing substance, a shining slime trail, when they're agitated. The function may be to distract predators. There's even a land snail that beams a
bluish light from near its head. When bothered, the snail flashes its light
off and on, perhaps to frighten the attacker.

Most bioluminescent animals live in the world's oceans, an environment that covers more than 70 percent of the earth's surface. Sunlight
rarely penetrates farther than 330 feet into the water. Since the average
depth to the oceans' bottoms is about 12,200 feet, about 97 percent of the
marine environment remains in perpetual darkness. Nonetheless, diverse
and highly specialized animals live in this vast region. Some of these—
such as deep-sea shrimps, many worms, jellyfish, sea anemones, corals,
and clams—give off light.

Some other bioluminescent animals, such as certain squid, swim around
between the surface waters and deep-sea areas. The jeweled squid, a spectacular bioluminescent mollusk with eight lights on the underside of its
body, glows ruby red in front, blue in the middle of the body, and white at
the rear. Some squid have luminous tentacles and arms; lights accent the
eyeballs of others.

Bioluminescence in squid probably serves several functions. It may

communicate species identity and possibly sex. Concentration of most of
the light-emitting areas on the underside of the body suggests that bioluminescence may also serve as camouflage. Sun- and moonlight shine
down on the squid, making its silhouette conspicuous from below. To
counteract this effect, some squid produce an intensity of light matching
that of the downwelling light—rendering the animal invisible to predators
below. Some squid have a small gland filled with luminescent bacteria near
the ink sac. The squid releases the glowing bacteria along with its ink, presumably as a way of startling or confusing predators.

Over 230 species of lantern fish have luminous organs running along
the sides of the head and down the length of the lower body. These small
marine fish, usually six inches or less, use their lights to attract prey and
to communicate with each other. Biologists use the characteristic patterns
of the fishes' luminous organs to distinguish among species, and perhaps
lantern fish recognize potential mates by the characteristic arrangement of
their lights. As with bioluminescent squid, the glowing light from below
matches the light intensity coming overhead from the sun or moon and
makes the fish invisible to predators beneath them.

Flashlight fish (*Photoblepharon*, meaning "eyelid light") also have multi-
purpose lights: an organ below each eye filled with continuously light-
emitting bacteria. Illumination coming from a fish is about as strong as
a weak flashlight. At night these marine fish venture out from their daytime retreats in dark caves and crevices in deep reefs to forage on small
crustaceans. James Morin and his colleagues worked with this fish in the
Red Sea in the 1970s and showed that they use their lights to see and lure
prey, escape or avoid predators, and to communicate with each other.

Flashlight fish exhibit several patterns of bioluminescence. One is in-
frequent blinking. A fish turns off its light, or blinks, by raising a black
eyelid-like structure over the light organ. Most bioluminescent animals
that flash their lights keep their lights off most of the time. In contrast,
the blink of flashlight fish is rapid, so the light is on most of the time.
The investigators suggested that in this behavioral mode, the light allows the fish to see and attract their crustacean prey that have well-
developed photoreceptors.

A second pattern is blink-and-run behavior, where the fish slowly
swims with the light on, then blinks the light and abruptly switches direction and swims faster. The next time it shines its light, the fish is in a
different location from its previous swimming course. This erratic swimming presumably helps to evade predators.

The fish may use several behavioral patterns to communicate with each other. Field observations revealed that pairs of fish defended territories along the reef against other flashlight fish. When an intruder approached, the female rapidly swam back and forth, then turned her light off, swam toward the intruding fish, and when she was next to the intruder she shone her light. The intruder swam off. The investigators carried out a laboratory experiment during which they discovered that two fish increased their blinking rate when they could see each other. When a black panel separated the fish, their blinking rates were ten per minute for fish A and fifty per minute for fish B. When the investigators removed the panel, these rates increased to forty and sixty per minute, respectively. Don't you wonder what the fish were "saying" to each other?

Humans, in their never-ending quest to exploit nature, have capitalized on these light organs. In 1922 an expert on bioluminescence from Princeton University recorded that fishermen of the Banda Islands in Indonesia remove the organs from flashlight fish and attach them to their lines above the hooks. The organs, which continue to shine for many hours, act as lures to attract other fish. Other Indonesian fishermen use flashlight fish as lures without killing them. They place a dozen or more flashlight fish in a perforated length of bamboo, which they suspend below their canoe. The light source acts as a magnet for other fish and can be used night after night.

AMONG ALL THE bioluminescent animals, from anemones to worms, most people favor the fireflies—perhaps because a meadow of dancing lights brings back childhood memories of magical summer evenings. Many of us collected fireflies in a glass jar. Under stress, our captives left their lights on, instead of flashing them. We couldn't quite read by these glowing lanterns, but they provided comfort in our dark tents.

Perhaps because of our fondness for fireflies, we associate these beetles with good fortune. Some people believe that if a firefly appears inside your house, a welcome visitor will come soon. Two fireflies in the house foretell a coming marriage, provided an unmarried woman lives there. More than two fireflies in the house means that cheerful company is on the way.

Glowing light—whether to communicate to others of the same species, attract prey, improve vision, or protect from enemies—is effective both for nocturnal terrestrial animals and for oceanic forms. Wouldn't it be grand if humans could communicate with each other at night by blinking bioluminescent body parts in Morse code!

RAPTUROUS AND RAPACIOUS REPTILES

The Turtle

The turtle lives twixt plated decks
Which practically conceal its sex.
I think it clever of the turtle
In such a fix to be so fertile.

OGDEN NASH, from *I Wouldn't Have Missed It*

Tactile communication is widespread throughout the animal kingdom and is especially common during courtship and mating (see scattered examples in chapter 1). We haven't looked at reptiles in this regard, however. Some reptiles include touch in their repertoire of amorous behaviors. But the touch isn't always a gentle one. Sometimes it's a bit rough, at least from our perspective.

OGDEN NASH HAD good reason to applaud the fertile turtle. As you can imagine, it's hard for a male turtle to mount even a slowly moving target. He must climb on from the rear and then balance his nearly flat or slightly concave plastron (lower shell) on the female's rounded carapace (upper shell). Though endowed with a penis unusually long for his body size, a male turtle can't use it unless the female stays still and cooperates. Many turtles use touch to get their lady loves in the mood.

Consider red-eared turtles, which live in quiet ponds with muddy bottoms and thick vegetation. We most often see these small (five to eight inches in length) brown or olive-green turtles with striking red ear stripes lazily basking on logs or masses of floating plants, but they're more active under the water surface.

Courtship in red-eared turtles involves an eager male lavishing attention on a passive and indifferent female. It begins when a male swims toward a female and sniffs her cloaca (the cavity into which the intestinal and genital-urinary tracts empty). If chemical cues assure him he has found the right species and sex, he faces the female and extends his front legs along either side of her head to signal his amorous intentions. He places his front feet, claws spread, over her eyes

and quivers his claws. She withdraws her head slightly into her shell and closes her eyes. The male rapidly, but gently, drums against her eyes with his claws—a behavior scientists call titillation. He performs anywhere from 24 to 125 bouts of titillation, each bout lasting a little longer than one and a half seconds. Eventually he slowly stops drumming and rests his front feet over her eyes momentarily before breaking contact. If the female is overwhelmed by his foreplay, he darts behind and mounts her.

And then there's the more aggressive side to turtle touch. Males of some species bite their mates' heads, shells, or legs to immobilize them. In addition to biting, male tortoises ram the females with the front of their plastrons. Eventually a receptive female lifts her rear and exposes her cloaca; the male mounts. A male wood turtle lunges, mounts, and tightly grips the female's shell on both sides of her head and tail. Then he shakes her side to side, sometimes for several hours. He bites her head, rubs his plastron across her carapace, and thumps their shells together—all this to convince her to cooperate.

EVEN WITHOUT FEET and claws to titillate with, snakes are tactile creatures and some go in for heavy petting. For many snakes, courtship begins when the male explores the female's body by flicking his tongue

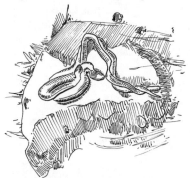 over her back and then rubbing his chin against her body. He may lie on top of her and twitch or send waves of muscle contractions along his body; he may tap or nudge her head or entwine his body around hers. The goal is to get the female to straighten her body. A male boa or python lies on top of a female and strokes and vibrates her back and sides with hard spurs (remnants of the pelvic girdle). Eventually, for many species, if all goes well and the female doesn't flee, the male searches for her cloaca with the tip of his tail. If she's receptive, she gapes her cloaca.

Courtship in western massasauga rattlesnakes lasts several hours, during which these sixteen- to forty-inch heavy-bodied snakes lie next to each other. The male rests his head and neck on the female's neck and head, and he loops his tail around hers. He strokes her at both ends, but not at the same time. For a while he rubs her head and neck with his chin, then he tightens the loop around her tail and massages the last two inches, up to her rattle. He strokes her head and neck again, and then back to her tail.

But all is not rapturous in the ophidian world of love. Some male snakes bite a female in her neck if she's coiled and not in the proper position to be courted. In some species, male neck bites enhance female receptivity and may signal the male's willingness to mate. If receptive, the female tolerates the bite. If not, she tries to escape. Either way, the male knows where he stands (so to speak). Some males even hold on to the gentler sex once courtship is over and while they're actually mating. Male Yuma kingsnakes grasp the females between their jaws while copulating, for up to two hours.

THE OTHER THREE groups of reptiles (lizards, crocodilians, and tuatara) use tactile stimulation to a lesser extent. Many male lizards lick females or rub their cloacas at the beginning of courtship, to detect chemical cues. Some male lizards nip and nudge their lovers, and some grip the females' tails in their jaws and bite. A female alligator usually indicates her interest by resting her snout on the male's head or snout. Both alligators then mutually rub heads and snouts. If the female is still receptive, they circle each other in the water, continue to rub heads, and blow bubbles from their mouths—rather playful, considering their size and strength!

The final group of reptiles are the tuatara: brown lizardlike creatures more closely related to the extinct dinosaurs than to modern-day lizards. A large male can be two feet long; females are a little smaller. Both sexes have crests of spines running down the centers of their backs. The only two species of tuatara alive today live on about thirty small islands off the coast of New Zealand.

Male tuatara initiate courtship by inflating their bodies, erecting the crests on their necks and backs, and approaching females in an ostentatious walk called a *stolzer Gang* (translated "proud walk"). While the male moves forward, he first pulls each leg slightly backward, then raises it, brings it forward, and then puts it down onto the substrate, as if to say, "Look at me. I'm big and handsome." There's also a tactile component to courtship. Sometimes, to get a reluctant female to cooperate, a male may lunge, gape widely, and bite her neck.

AND WHAT ABOUT the dinosaurs, those reptiles of yesteryear? Did they use tactile stimulation to communicate during courtship? And if they did, were they rapturous or rapacious? Did *Apatosaurus* males and females entwine their twenty-foot necks and waltz among the tree ferns? Or did the thirty-ton male sit on the female and subdue her with his claw? Did male and female *Tyrannosaurus rex* clasp their tiny two-fingered front feet together in some sort of courtship ritual? Or did the males live up to their name of "tyrant lizard" and sink their sharp six-inch teeth into females' necks to force them into submission? What about *Stegosaurus?* How do you mount a bony-plated giant with three-foot-long spikes on her tail? Did males subdue females by cautious caresses or by force? We'll never know.

SMELLING IS UNDERSTANDING

. . . an antelope passes along a certain plain, eats at one place, drinks at another, lies down at a third, is pursued by a wolf for half a mile, when the wolf gives up the unequal race, and the antelope escapes at his ease. A second antelope comes along. The foot scent from the interdigital glands marks the course of his relative as clearly for him as the track in the snow would for us. Its strength tells him somewhat of the time elapsed since it was made, and its individuality tells him whether his predecessor was a stranger or a personal friend, just as surely as a dog can tell his master's track. The frequency of the tracks shows that the first one was not in haste, and the hock scent, exuded on the plants or ground where he lay down, informs the second one of the action. At the place where the wolf was sighted, the sudden diffusion of the rump musk on the surrounding sage-brush will be perceptible to the newcomer for hours afterward. The wide gaps between the traces of foot scent now attest the speed of the fugitive, and the cause of it is clearly read when the wolf trail joins on.

ERNEST THOMPSON SETON, Annual Report of the
Smithsonian Institution 1901, from *Ernest Thompson Seton's America*

Don't be put off by the long quote above. Read it for a unique insight into the importance of scent for communication in antelope. Ernest Thompson Seton, an astute naturalist (and a founder of the Boy Scouts of America), wrote this scenario over a century ago.

The scents that Seton wrote about are chemical signals called pheromones (from the Greek words *pherein* and *horman,* meaning "to transfer" and "to excite"). Pheromones are secreted by one animal and picked up by another individual of the same species; they cause a behavioral or physiological response in the receiver.

These chemicals can be dispersed through the air or water, by surface contact, or directly between individuals. Many organisms, from slime molds to primates, use them. (Chemicals that convey information between individuals of different species are called allelochemics. We'll consider only pheromones here.)

Seton's quote describes the various bits of information an antelope receives from another's foot scents. At the end of this chapter, we'll return to the idea that an animal can tell whether another individual is familiar or unfamiliar. Here we'll look at other types of information conveyed by pheromones.

ONE MAIN CATEGORY of pheromones, called primer pheromones, induce relatively long-lasting physiological changes in the recipients. The receiver becomes "primed" for a new set of behaviors. A classic example of primers are certain pheromones secreted by queen bees.

Have you ever wondered about the origin and significance of the expression "queen bee"? A honeybee queen maintains stability and cohesion in the colony just by her presence, and she's the focus of constant attention. When the queen is stationary, eight or more attendants, her "court," face her in a circle and offer her food, palpate her with their antennae, and lick her. Workers wait on her hand and foot: they forage for pollen and nectar, rear the brood, and build the honeycomb for storing food and raising the young bees.

A queen honeybee exerts such power because she produces pheromones that attract workers to her and make her irresistible. She also secretes a pheromone that inhibits ovary development in the workers, preventing all other females in the hive from becoming sexually mature. She, then, is the only female who can reproduce.

Another primer influence is called the "Bruce effect," after Hilda Bruce, the scientist who first reported its existence in laboratory mice. The Bruce effect happens when a recently impregnated female mouse encounters a male whose urine smells sufficiently different from her mate's urine. Her fertilized egg fails to implant, and she comes into estrus again within a week. Through his pheromones, the second male has caused the female to become receptive again—and he will gladly provide the service. Male-induced termination of pregnancy also occurs in prairie deer mice and some voles. Biologists are still puzzling over how to interpret this phenomenon in evolutionary terms. Obviously, the newcomer male benefits, but what's in it for the female?

PHEROMONES OF THE second main category, called releasers, act directly on the recipient's central nervous system, stimulating more or less immediate behavioral responses in the receiver. We'll consider four types of releaser pheromones: alarm, home range, funeral, and trail.

After detecting danger, some animals secrete pheromones that function as alarm signals and incite other members of the group into defensive behavior. Termites, ants, bees, and wasps are classic alarm pheromone producers, so let's return to honeybees. Honeybee workers guard their hive's entrance from honey-stealing intruders. When a guard stings an intruder, her barbed stinger gets caught in the victim. When the guard flies away, she leaves behind her stinger and attached poison gland, exposing alarm pheromones. These chemicals evaporate rapidly and attract nest mates, ready and willing to execute a mass assault to protect the hive with their own stingers—and their lives.

A second class of releasers, called home range pheromones, are deposited by animals when they mark the environment to communicate their presence or dominance to other individuals. We all know about male dogs and fire hydrants, but if you took a giant panda for a walk, he (or she) would also mark the path.

Giant pandas from China don't often run into each other, except during the spring mating season. Males have large home ranges (the area through which an animal habitually roams) that overlap with those of other individuals. Pandas mark sites along their frequently traveled routes with urine and secretions from their anogenital glands. This communication system allows them to locate chemicals of other individuals and to deposit their own chemicals in areas of overlap. Generally, one large dominant male rules a given area. Subadults avoid competition and aggression with dominant males by avoiding these areas.

A subadult distinguishes between an adult and another subadult by smell. It's not that simple, though, and not all adults are equally dominant. Pandas convey critical information about their size and dominance through the height at which they deposit their odors. Giant pandas mark in four different postures. In the "squat" position, a panda marks the ground. In the other three, the animal marks a vertical surface such as a rock or tree: In the "reverse" position, the panda backs up to a vertical surface; in the "legcock" posture, it raises one leg; and in the "handstand" posture, it extends its body and lifts its hind legs off the ground. Which posture a panda uses depends in large part on its age and sex. Although all pandas squat, females and subadults use this position most often. Subadults occasionally use the reverse and legcock postures, and females sometimes legcock. Adult males use these latter two positions often, and only they use the handstand posture.

The type of scent and height at which the scent is deposited vary with the posture. When squatting, a panda leaves either glandular secretions, urine, or sometimes both on the ground. An animal in the reverse or leg-cock posture deposits primarily glandular secretions about eighteen inches above the ground. Males in the handstand position squirt mostly urine, and it lands about three feet above the ground.

Angela White, of the Zoological Society of San Diego, and her colleagues experimented with twenty-eight pandas in the Wolong Nature Reserve in Sichuan, China, to determine whether pandas respond differently to scents deposited at different heights. They placed male urine and male and female glandular secretion at different heights on the enclosure walls and on the floor to mimic marking postures and then recorded the pandas' behavioral responses (sniffing, deep inhalation, licking, and rubbing the anogenital gland against the area or depositing a small amount of urine) to the different artificially marked sites. They also measured how far each panda stayed from the odor site.

The pandas spent more time investigating odors higher up on the enclosure walls than the ones lower down, suggesting they were more interested in individuals that mark in high places. Although they spent more time investigating odors on high, the experimental subjects (especially subadult males) stayed farther away from male urine placed to mimic the handstand posture than from male urine on the ground. By marking in an elevated posture, a panda apparently communicates the message: "I'm big and dominant."

Giant pandas aren't the only mammals that literally do handstands to leave their scent in high places. Dwarf mongooses, giant rats, foxes, badgers, and lemurs also stand on their front feet, extend their bodies, and lift their rears to deposit their odors as high as possible. Perhaps, like pandas, by aiming high they provide cues to their body size and competitive ability.

Funeral pheromones released by dead ants are a third class of releasers. When a nest mate detects this odor, it picks up its dead comrade and carries it to a refuse pile outside the nest. As you can imagine, necrophoresis (the technical word for removal of dead nest mates) improves hygiene in the nest. The urge to keep death out of the nest is so strong that if you smear some funeral pheromone on a small inanimate object or on a live worker, the industrious undertakers will pick up the object or ant and dump it in the garbage pile. And what about the unfortunate live ant that has been dumped? It cleans itself off and returns to its duties in the nest, seemingly unfazed.

Trail pheromones, a fourth class of releasers, are laid down by individuals and then detected by others following the trail. Walt Disney's animated movie *A Bug's Life* begins with the ants carrying food back to their nest when a leaf falls over their invisible path. There's instant panic until the "trained professionals" instruct the ants to walk around the leaf back to "the line"—in scientific terms, the pheromone trail.

In real life, when a foraging worker ant finds food, it scurries back to the nest, all the while depositing trail pheromone, which directs nest mates to the source. As they return to the nest with food, the recruited workers reinforce the trail with their own trail pheromones. Once all the food has been taken, the workers returning to the nest no longer reinforce the scent and the trail evaporates.

Eastern tent caterpillars also lay down scent trails. These caterpillars live gregariously in colonies in their host trees, primarily apple and cherry trees, where they build white tentlike shelters from strands of silk. Although the adult moth is a rather blah brown, the caterpillars are striking: dark gray with blue and white or yellow spots along their sides, a yellow racing stripe running down the center of the back, and a covering of wispy beige or cream hairs.

Tent caterpillars generally leave the safety of their tent and forage soon after sunrise, early afternoon, dusk, and early morning before sunrise; they return to the tent after each feeding bout. While moving over branches, tent caterpillars secrete strands of silk. As they leave the tent in search of young tender leaves, they brush the tips of their abdomens along their silk

trails, depositing exploratory pheromones. When successful foragers return home to the tent, they deposit different chemicals along these same trails—which then become recruitment trails. Hungry tent mates react excitedly to these recruitment chemicals, follow the trails to the source, and feast.

Terrence Fitzgerald and his colleagues have found that these caterpillars don't recruit tent mates to sparse feeding sites. Instead, they return along their exploratory trails holding their abdomen tips above, not on, the branch surfaces. The caterpillars also take into account food quality. When the investigators gave caterpillars young leaves of aspen, blueberry, willow, and other nonhost trees, the caterpillars ate them but deposited less recruitment pheromone than caterpillars fed on cherry leaves. In another experiment, they fed caterpillars either tender young host plant leaves or older host plant leaves, which are tougher, drier, and have less nutrient content. The two groups ate about the same amount, but caterpillars fed on the higher-quality young leaves deposited lots more recruitment pheromone.

AS YOU CAN SEE, some animals couldn't function without chemical communication. Ants' nests would pile up with corpses. Other members of the hive might become sexually mature and compete with the queen honeybee. And tent caterpillars would be on their own to find juicy leaves to devour.

LOVE POTION NUMBER NINE

I took my troubles down to Madame Ruth
You know that gypsy with the gold-capped tooth
She's got a pad down on Thirty-Fourth and Vine
Sellin' little bottles of Love Potion Number Nine

I told her that I was a flop with chicks
I've been this way since 1956
She looked at my palm and she made a magic sign
She said, "What you need is Love Potion Number Nine"

JERRY LEIBER and MIKE STOLLER, "Love Potion Number Nine"

For millennia, men and women have sought love potions, aphrodisiacs, and sex attractants. Some, like the poor misguided soul in the song, drink love potions in hopes of increasing their success in love. Others eat (or feed to others) chocolate, oysters, ginseng, or powdered rhinoceros horn to increase sexual desire. And some people rub perfumes made from animal, plant, and synthetic substances onto their skin—a custom that dates back at least to the ancient Egyptians three thousand years ago—to increase sex appeal and attractiveness.

Many animals naturally produce scents that serve as sex pheromones. One type of sex pheromone, the sexual attractants, are often broadcast over long distances. Generally, the female releases these chemicals to advertise her presence and to lure in males. Anyone who has ever had a female dog or cat in heat knows the powerful attraction the female's urine holds for the neighborhood bachelors. Aphrodisiacs, the other category of sex pheromones, are used for short-range communication. Males release aphrodisiacs to grease the wheels of courtship or to persuade those females within reach to mate.

The pheromone disparlure is a well-studied sex attractant that a female gypsy moth releases from a scent gland on her abdomen. Too heavy with eggs to fly, the creamy-white female flutters her wings to waft her alluring scent into the air. From the time she emerges from her cocoon and releases her perfume until she dies seven to ten days later, her sole purpose is to get her one hundred to eight hundred eggs fertilized and then to deposit the amber-colored egg mass. The brown strong-flying males use their large feathery antennae, covered with odor-receptor hairs, to detect disparlure up to seven miles away. They can respond to 4 millionths of a billionth of an ounce of scent. Top that, Coco Chanel.

After she tosses out her lure, a female gypsy moth waits in a characteristic position: head upward, abdomen lowered, and wings slightly spread. Lustful males zero in. If several males arrive at the same time, they thwack each other with their wings until all but one throws in the towel. After copulation, the female shakes off the male and releases a counter-pheromone, a different chemical that neutralizes any trace of her original perfume. Then, in the calm following the onslaught of scent-maddened males, she searches for a spot to lay her eggs.

Gypsy moth caterpillars are serious pests of forest, shade, and fruit trees in the northeastern United States. A mature two-inch caterpillar—brownish black with tufts of hairs and five pairs of blue spots and six pairs of red spots on its back—can devour a square foot of leaves each day. During large outbreaks, ravenous caterpillars completely strip trees of all their leaves, defoliating over a million acres of northeastern woodlands in a year.

Over the past half century, we've developed chemical, biological, and microbial methods to kill these pests. One innovative method uses a synthetic form of the moths' sex attractant to prevent males from finding females. Horny males are lured to scent-laced traps, where they're killed. Alternatively, biological control experts can put out enough synthetic disparlure that males become confused and can't find real-life females throwing out scent. Unfortunately, when female densities are high, no amount of synthetic disparlure can divert males from real-life females.

LET'S TURN TO some very different animals: red-sided garter snakes from Manitoba, in western Canada. These snakes live farther north than any other reptile in North America. Winters there are harsh for snakes, sometimes with continuous snow cover from late September through May. Temperatures may reach as low as $-50°F$. The snakes escape these extreme conditions about nine months of the year by aggregating and remaining dormant in underground dens. Sometimes more than ten thousand snakes gather in one den, the highest concentrations of snakes in the world. Once temperatures warm up in late April to mid-May, the snakes are ready for action. They have only about three warm months during which they can mate, feed, and give birth. Then it's back underground for another long nap.

Most of the males emerge from the den early in the breeding season, and they stay in the area for about a month. Females, in contrast, emerge singly or in small groups scattered throughout the month. For this reason, the sex ratio is strongly skewed toward males. On any given day, females are few and far between, leading to intense competition among males for mates.

Males emerge from winter dormancy with active sperm, and they lose no time in finding potential mates. Finding females is the easy part. Female red-sided garter snakes release a pheromone from their skin lipids— a chemical that makes them so attractive to males that a female sparks a frenzied response among nearby males. The result might be a mating ball of up to a hundred snakes. Each male writhes and jostles to get into a good position for that one female. If she can, a female avoids this harassment by quickly slithering away from the den before she gets trapped. If she successfully escapes the horde of males near the den entrance, she'll be intercepted by a small group of males or a single male that finds her by following her chemical trail. Over 95 percent of the females mate within one day of emerging from the den. When a lucky male successfully copulates, his sperm is accompanied by a male pheromone that renders the female unattractive to other males. He also blocks her cloaca with a gelatinous plug, which helps to ensure she won't double-time him.

If given a choice, though, male red-sided garter snakes don't court just any female. They prefer their lovers "robust," as revealed by recent experiments carried out by Michael Lemaster and Robert Mason. The experimenters first put males into an arena with two females: one small (less than twenty inches) and one large (more than twenty-four inches). Significantly more males courted the big mamas than the puny females.

In a second experiment, the investigators tested whether male garter snakes can distinguish the size of females based on smell alone. They placed ten males into an arena with either a small female, a large female, filter paper containing skin lipid from a small female (they rubbed the backs of small females onto wet filter paper), or wet filter paper containing skin lipid from a large female. After five minutes, the investigators recorded the number of males exhibiting courting behavior: rapidly tongue flicking and chin rubbing either the female or the filter paper, and aligning the male's body alongside the female or the edge of the filter paper.

The big mamas and their scents won out. Significantly more males courted the large females and the filter paper containing large-female scent than the small females or filter paper containing small-female scent. Furthermore, as many males courted the paper as the female snakes! Lemaster and Mason concluded that males can discriminate between large and small females based solely on the chemical cues in the females' skin. They don't need visual, tactile, or behavioral cues. The sex-attractant pheromone of larger females has a different chemical composition than that of smaller females, and males cue in on that.

SOME SALAMANDERS USE aphrodisiacs. During courtship, salamanders of the family Plethodontidae (lungless salamanders) rely heavily on pheromones produced by glands in the chin and cloacal regions. A female plethodontid salamander signals her receptivity to a courting male by entering into a tail-straddling walk: she places her chin over the base of the male's tail and takes in the aroma of his cloacal aphrodisiac. The chem-

icals travel by capillary action up the female's nasolabial grooves (grooves that run from her upper lip to the corners of her nostrils) to specialized organs in her nose. In tandem, they amble forward along the ground or bottom of the pond or stream, depending on the species. While walking, the male deposits a spermatophore (packet of sperm) onto the substrate. As she passes over it, the female vacuums up the spermatophore into her cloaca.

If the female isn't cooperative and needs extra persuasion during the tail-straddling walk, the male turns around and transfers aphrodisiac to her from his chin gland. In some species, the glands are large and padlike. These males deliver their pheromones by slapping their chins against the females' snouts—a direct hit, so the nasolabial grooves can transfer the chemicals to the sensory organs in the nose. Males from species with smaller chin glands deliver their scents by biting or by raking their teeth over the females' skin and then pressing their glands against the broken skin. Either way, they deliver the pheromone directly into the females' bloodstream. Males of several species take the direct approach: they bite into their mates, sometimes hanging on for several hours, and inject their aphrodisiacs.

AMONG MAMMALS, some bats are master perfumers. The greater sac-winged bat is a small (about two and a half inches maximum length) dark brownish-black bat that lives in colonies of up to sixty individuals, in the hollows of large trees in Central and South America. Within a colony, males defend territories and maintain harems of up to seven females. Males blend their own aphrodisiacs from urine and odiferous glandular secretions, and they store their perfume in saclike organs located on their wing membranes. Then they hover in front of roosting females within their territories, fanning their "come-on" scent.

A male sac-winged bat spends up to an hour every afternoon cleaning, moistening, and refilling his paired wing sacs. After licking his sacs, he bends his head toward his privates, takes one or several drops of urine in his mouth, and licks the urine onto one of his wing sacs. Then back for more urine. After spending about seven minutes transferring urine to his wing sacs, he rests for several minutes. Then, for about the next twenty-two minutes, he presses his throat to his penis and transfers small droplets of glandular secretion to the wing sacs. After resting a few seconds more, he opens his mouth and squeezes out droplets from his throat gland and transfers them to his sacs. Then back to his genital area, to repeat the sequence. After blending his chemical concoction, a male usually either sleeps for a while or grooms himself. He'll fan his perfume of the day for his harem females before they leave the tree at dusk and after they reenter at dawn.

AND WHAT ABOUT humans? Do we have natural pheromones that serve as sex attractants or aphrodisiacs? Despite the fact that people frequently say, "There's just no chemistry between us," or, "The chemistry was incredible!" the scientific evidence is inconclusive.

Consider armpit sweat. Men and women have different amounts of certain types of bacteria in their armpits, causing their sweat to smell different. The fact that armpit scent glands only become active at puberty and underarm hair appears simultaneously suggests that armpit scent may be a pheromone involved in sexual communication. (Hair effectively spreads odor because of its large evaporative surface area.) Investigators have found three steroids in human sweat that may function as pheromones. These steroids (androstenone, androstenol, and androstadienone) are manufactured by the testes, ovaries, and adrenal glands. By themselves, these steroids don't smell like much, but once the armpit bacteria work on them, they smell musky.

Scientists have carried out experiments to determine if women are affected by the odor of androstenol or androstenone in male sweat. Some studies have shown that women find male subjects with these chemicals more attractive. Other studies, though, have shown opposite or no effects.

Despite variable experimental results, these steroids are aggressively marketed as sex attractants. Some products have a high pheromone concentrate: you rub these into your skin or spray them on. Others are herbal supplement pills that purportedly help to stimulate the production and release of men's own natural pheromones, the implication being that the buyer had previously been short on sexy scents. All these pheromone colognes, perfumes, and pills supposedly make the user more approachable and more desirable to the opposite sex, encourage thoughts of romance, and induce receptiveness to the user's advances. But do they really work? Perhaps, but maybe it's the power of suggestion or increased self-confidence. Whichever the case, we'll probably continue the search for love potions, aphrodisiacs, and sex attractants as long as we're around.

IT PAYS TO BE NEIGHBORLY

Good fences make good neighbors.

ROBERT FROST, "Mending Wall"

Territorial animals maintain an area around themselves in which they feel secure. They stay in one area (in scientific jargon, they exhibit site fidelity) and defend it from others of the same species. Animals use visual, acoustic, and chemical signals to stake out territories and to communicate their ownership to others. Many lizards patrol their territory boundaries, bob their heads up and down, and do push-ups. They're saying: "Stay out." Birds, crickets, and frogs sing, chirp, and croak. Mammals mark with scents, including urine, feces, and musk.

Owning a territory ensures the owner exclusive use of the area's resources, including nest sites and food. But, like everything else in life, where there are benefits, there are also costs. An animal may spend considerable time and energy defending its territory, expose itself to predators, and risk injury during fights with intruders. Because of the costs involved, a territory owner shouldn't go ballistic every time an intruder oversteps the boundary.

The intruder might be a neighbor with its own territory, an individual with whom he or she has worked out a mutual boundary agreement; a minor infringement may not be worth hostile exchange. Or the intruder might be a stranger skulking about looking for an area of its own. Because a stranger is generally a greater threat than a neighbor, a territory owner should distinguish between the two and react differently.

Here we'll look at several animals that recognize neighbors and respond less aggressively to them than to strangers, a behavior known as the "dear enemy phenomenon" (so-called because neighbors start out as foes). By recognizing neighbors and responding less aggressively to them, animals reduce their cost of defending territories. This behavior occurs mainly in animals whose territories include both critical food supplies and breeding sites.

MOST MALE BIRDS that defend territories advertise their boundaries by singing. Some species—such as American robins, cardinals, ovenbirds, white-throated sparrows, great tits, and others—recognize their neighbors' songs. If you play a tape recording of a neighbor's song, the territorial resident barely reacts; if you play a recording of a stranger's song, the resident male becomes agitated and upset. He may approach the loudspeaker, hop up and down with ruffled feathers, and castigate the machine vocally.

What about birds that sing more than one type of song? Can an individual sort through song repertoires and distinguish between those of neighbors versus strangers? Experiments to address this question were carried out in New Brunswick, Canada, with veeries (a thrush about seven inches long). Veeries sing a flutelike "veer, veer, veer, veer," a rolling series of notes dropping down the scale. Males generally have two to four types of songs, and they produce several versions of each type by changing the number of syllables. When the investigators played tapes of neighbors' songs versus strangers' songs to experimental birds on their woodland territories, the birds responded more strongly to strangers' songs. They approached the loudspeaker more quickly, spent more time near the noisy contraption, and sang more in an effort to oust the perceived intruder.

SOME AMPHIBIANS ARE more aggressive toward strangers than to neighbors. Red-backed salamanders, small (three and a half to five inches in length) inhabitants of the leaf litter in woodlands throughout the northeast United States, establish territories under and around logs and rocks that serve as refuge sites. Robert Jaeger and his collaborators have found that red-backed salamanders mark their territory boundaries with feces, and they defend them from other red-backed salamanders by threat displays and by biting. The threat display is a "look big" posture where the animal extends all four legs, raises its head and entire trunk off the ground, and faces the opponent. When biting, a salamander usually aims for the snout or tail—two particularly vulnerable areas. Damage to the snout may reduce smelling ability, and a bite to the tail may cause the salamander to self-amputate (autotomize) its tail. Experiments reveal that adult red-backed salamanders distinguish between neighbors and strangers by the smell of their feces, and they're less aggressive and more submissive toward neighbors.

In these salamanders, "dear enemy" recognition even extends into the courtship season. Laboratory experiments reveal that males spend more time in threatening postures when exposed to unfamiliar gravid (full of ripe eggs) females and more time touching familiar gravid females, possibly indicating their willingness to court. Females spend more time flat against the ground in a submissive posture with familiar males than with unfamiliar males. In the case of red-backed salamanders, familiarity doesn't breed contempt—familiarity breeds.

Why should a male red-backed salamander threaten unfamiliar females? Additional work is needed to answer this question, but perhaps it has to do with a finite quantity of food in a male's territory. Perhaps there's a limit to the number of intruders he will allow. Neighboring females may be okay because there's a good chance he can eventually mate with them. Strangers are not okay because he may never see them again.

SOME TERRITORIAL MAMMALS, such as beavers, also respond less aggressively to neighbors. Like the red-backed salamanders, they distinguish neighbors from strangers by smell. Beavers usually live in

family groups consisting of an adult male and female, kits, yearlings, and sometimes two-year-olds. Adults mark their territories by depositing castoreum (a smelly substance found in their castor glands) and secretion from their anal glands onto small mounds of mud, grass, or sticks on land, but close to the water.

Experiments with Eurasian beavers in Norway reveal that adults spend much more time sniffing at mounds covered with strangers' scents and more time acting aggressively at these mounds than at mounds covered with neighbors' scents. They stand on their hind feet on stranger mounds, paw and scratch the mound, and then dump a pile of mud either at the side or on top of the mound. As if defiant, they mark the mound with their own castoreum, anal secretion, or both.

Beavers are monogamous, and an adult pair lives in the same territory for many years. Thus, from the standpoint of time and energy, it pays for them to recognize their neighbors' scent and to tolerate their occasional incursions and close proximity. Beavers are feisty when provoked, and their battles can lead to serious injury and even death—another reason neighbors shouldn't fight with each other.

BIRDS RECOGNIZE NEIGHBORS by song, and red-backed salamanders and beavers do it by smell. Lizards use vision. One discriminating lizard is the Augrabies flat lizard, which lives on rocky outcrops in Northern Cape Province, South Africa. Males set up territories on rocks and defend them from male intruders. A male elevates one side of his body to flash an otherwise concealed patch of yellow or orange (or a combination of colors) on his belly. By flashing this "status-signaling badge," a male conveys aggression toward the intruder—a warning to get out. If that doesn't work, he may chase the trespasser. Males strut about and patrol their territory boundaries, so they have plenty of opportunity to get to know each other as neighbors. In areas where lizard densities are high, territories are at a premium and some males are left out—remember playing musical chairs as a kid? Males without territories wander about in search of uncharted land, crossing boundary lines and causing territorial disputes.

Field experiments reveal that territorial male Augrabies flat lizards allow neighbors to approach much closer than non-neighbors before they challenge, and they chase non-neighbors significantly farther away. Interactions between non-neighbors are more aggressive and more likely to culminate in combat that includes biting. Clearly, these lizards conserve time and energy and take fewer risks by being neighborly.

CLOSER TO HOME . . . you might grumble when the neighbor kids climb over your fence and tromp across your manicured lawn to retrieve their soccer ball, but you'll likely yell louder if you spot a bunch of strangers doing the same.

Parting Thoughts

WE'VE EXPLORED a heterogeneous assortment of animal lifestyles. I hope you found them as amazing as I do: sex-changing clownfish; bowerbirds that decorate their boudoirs; tadpole-toting hip-pocket frogs; penis-gnawing hermaphroditic giant slugs; mites that die before they're born; dung-rolling scarabs; toe-wiggling Pac-Man frogs; intestine-ejecting sea cucumbers; boxer crabs that protect themselves with venomous sea anemones; and assassin bugs that pile prey corpses onto their backs.

Even more amazing to me is that diverse and unrelated animals share some of these bizarre behaviors—testimony that these natural histories are successful ways of living. Male sow bugs, giant cuttlefish, and bluegill sunfish masquerade as females and sneak copulations. Humans, wood turtles, and American robins thump the ground to lure earthworms to the surface. Male European nursery web spiders, dance flies, and ospreys offer gifts of food in return for sex. Fireflies, flashlight fish, and jeweled squid communicate by shining lights.

Yet even with the numerous examples I've mentioned, I've barely scratched the surface of the astounding variety of unusual animal natural histories. Zoologists could write dozens of tomes on the subject without ever repeating examples. I often wonder what other natural histories are out there yet to be discovered. You don't need a PhD or even professional training in zoology to observe and appreciate animal natural histories. All it takes is a keen eye and patience . . . though sometimes binoculars help. There's nothing quite like the thrill of discovery—whether a first for you, or for science.

Appendix: Scientific Classification

THE CLASSIFICATION SYSTEM we use today dates back to 1753, when the Swedish naturalist Carolus Linnaeus developed a hierarchical system of seven main levels. Kingdom is the broadest level; species is the narrowest. Organisms within a group at a given level are more similar to each other than they are to organisms belonging to a different group in that level. The basic classification is as follows, using two invertebrates and two amphibians as examples. Note that the invertebrates diverge after the kingdom level, whereas the amphibians share down to the same class, then diverge at the order level.

	pork tapeworm	black widow spider	strawberry poison dart frog	red-backed salamander
Kingdom	Animalia	Animalia	Animalia	Animalia
Phylum	Platyhelminthes	Arthropoda	Chordata	Chordata
Class	Cestoda	Arachnida	Amphibia	Amphibia
Order	Cyclophyllidea	Araneae	Anura	Caudata
Family	Taeniidae	Theridiidae	Dendrobatidae	Plethodontidae
Genus	*Taenia*	*Latrodectus*	*Dendrobates*	*Plethodon*
Species	*solium*	*mactans*	*pumilio*	*cinereus*

References Consulted
and Suggested Reading

Rampant Machismo

Attenborough, D. *The Trials of Life: A Natural History of Animal Behavior.* Boston: Little, Brown, 1990.

Coco, C. *Secrets of the Harem.* New York: Vendome Press, 1997.

Cox, C. R., and B. J. Le Boeuf. "Female Incitation of Male Competition: A Mechanism in Sexual Selection." *American Naturalist* 111 (1977): 317–35.

Halliday, T. *Sexual Strategy.* Chicago: University of Chicago Press, 1980.

Le Boeuf, B. J. "Male-Male Competition and Reproductive Success in Elephant Seals." *American Zoologist* 14 (1974): 163–76.

Le Boeuf, B. J., and R. M. Laws, eds. *Elephant Seals: Population Ecology, Behavior, and Physiology.* Berkeley: University of California Press, 1994.

Thomas, J. W., and D. E. Toweill, compilers and eds. *Elk of North America: Ecology and Management.* Harrisburg, PA: Stackpole Books, 1982.

Come Up and See My Etchings

Alcock, J. *The Kookaburras' Song: Exploring Animal Behavior in Australia.* Tucson: University of Arizona Press, 1988.

Borgia, G. "Why Do Bowerbirds Build Bowers?" *American Scientist* 83 (1995): 542–47.

Borgia, G., S. G. Pruett-Jones, and M. A. Pruett-Jones. "The Evolution of Bower-Building and the Assessment of Male Quality." *Zeitschrift fur Tierpsychologie* 67 (1985): 225–36.

Gilliard, E. T. *Birds of Paradise and Bower Birds.* London: Weidenfeld and Nicolson, 1969.

Marshall, A. J. *Bower-Birds: Their Displays and Breeding Cycles.* London: Oxford University Press, 1954.

Pruett-Jones, M., and S. Pruett-Jones. "The Bowerbird's Labor of Love." *Natural History* 92 (September 1983): 48–55.

Flying Casanovas. NOVA video. David Attenborough. 2001.

Sneakers and Deceivers

Arnold, S. J. "Sexual Behavior, Sexual Interference and Sexual Defense in the Salamanders *Ambystoma maculatum, Ambystoma tigrinum* and *Plethodon jordani.*" *Zeitschrift fur Tierpsychologie* 42 (1976): 247–300.

Barlow, G. W. "Social Behavior of a South American Leaf Fish, *Polycentrus schomburgkii,* with an Account of Recurring Pseudofemale Behavior." *American Midland Naturalist* 78 (1967): 215–34.

Birkhead, T., and A. Moller. "Female Control of Paternity." *Trends in Ecology and Evolution* 8 (1993): 100–104.

Bulfinch, T. *Bulfinch's Mythology*. New York: Gramercy Books, 1979.

Gross, M. R. "Sneakers, Satellites and Parentals: Polymorphic Mating Strategies in North American Sunfishes." *Zeitschrift fur Tierpsychologie* 60 (1982): 1–26.

Hall, K. C., and R. T. Hanlon. "Principal Features of the Mating System of a Large Spawning Aggregation of the Giant Australian Cuttlefish *Sepia apama* (Mollusca: Cephalopoda)." *Marine Biology* 140 (2002): 533–45.

Lance, S. L., and K. D. Wells. "Are Spring Peeper Satellite Males Physiologically Inferior to Calling Males?" *Copeia* (1993): 1162–66.

Shuster, S. M. "Alternative Reproductive Behaviors: Three Discrete Male Morphs in *Paracerceis sculpta*, an Intertidal Isopod from the Northern Gulf of California." *Journal of Crustacean Biology* 7 (1987): 318–27.

Shuster, S. M., and M. J. Wade. *Mating Systems and Strategies*. Princeton, NJ: Princeton University Press, 2003.

Headless Males Make Great Lovers

Andrade, M. C. B. "Sexual Selection for Male Sacrifice in the Australian Redback Spider." *Science* 271 (1996): 70–72.

Austad, S. N., and R. Thornhill. "This Bug's for You." *Natural History* 100 (December 1991): 44–48.

Buskirk, R. E., C. Frohlich, and K. G. Ross. "The Natural Selection of Sexual Cannibalism." *American Naturalist* 123 (1984): 612–25.

Elgar, M. A. "Sexual Cannibalism in Spiders and Other Invertebrates." In *Cannibalism: Ecology and Evolution among Diverse Taxa*, ed. M. A. Elgar and B. J. Crespi, pp. 128–55. Oxford: Oxford University Press, 1992.

Elgar, M. A., and D. R. Nash. "Sexual Cannibalism in the Garden Spider *Araneus diadematus*." *Animal Behaviour* 36 (1988): 1511–17.

Forster, L. M. "The Stereotyped Behaviour of Sexual Cannibalism in *Latrodectus hasselti* Thorell (Araneae: Theridiidae), the Australian Redback Spider." *Australian Journal of Zoology* 40 (1992): 1–11.

Maxwell, M. R. "Does a Single Meal Affect Female Reproductive Output in the Sexually Cannibalistic Praying Mantid *Iris oratoria?*" *Ecological Entomology* 25 (2000): 54–62.

Nash, O. *Custard and Company*, selected and illustrated by Quentin Blake. Boston: Little, Brown, 1980.

Polis, G. A. "The Evolution and Dynamics of Intraspecific Predation." *Annual Review of Ecology and Systematics* 12 (1981): 225–51.

———. "The Unkindest Sting of All." *Natural History* 98 (July 1989): 34–38.

Trading Food for Sex

Coe, S. D. *America's First Cuisines*. Austin: University of Texas Press, 1994.

Engqvist, L., and K. P. Saver. "Amorous Scorpionflies: Causes and Consequences of the Long Pairing Prelude of *Panorpa cognata*." *Animal Behaviour* 63 (2002): 667–75.

Fuller, L. K. *Chocolate Fads, Folklore, & Fantasies: 1,000+ Chunks of Chocolate Information.* New York: Haworth Press, 1994.

Gwynne, D. T. "Courtship Feeding Increases Female Reproductive Success in Bush-crickets." *Nature* 307 (1984): 361–63.

Mougeot, F., J.-C. Thibault, and V. Bretagnolle. "Effects of Territorial Intrusions, Court-ship Feedings and Mate Fidelity on the Copulation Behaviour of the Osprey." *Animal Behaviour* 64 (2002): 759–69.

Nisbet, I. C. T. "Courtship-Feeding and Clutch Size in Common Terns *Sterna hirundo.*" In *Evolutionary Ecology,* ed. B. Stonehouse and C. Perrins, pp. 101–9. London: University Park Press, 1977.

Simmons, L. W., et al. "Nuptial Feeding by Male Bushcrickets: An Indicator of Male Quality?" *Behavioral Ecology* 10 (1999): 263–69.

Tasker, C. R., and J. A. Mills. "A Functional Analysis of Courtship Feeding in the Red-Billed Gull, *Larus novaehollandiae scopulinus.*" *Behaviour* 77 (1981): 222–41.

Thornhill, R. "Adaptive Female-Mimicking Behavior in a Scorpionfly." *Science* 205 (1979): 412–14.

———. "Sexual Selection and Nuptial Feeding Behavior in *Bittacus apicalis* (Insecta: Mecoptera)." *American Naturalist* 110 (1976): 529–48.

Thornhill, R., and J. Alcock. *The Evolution of Insect Mating Systems.* Cambridge, MA: Harvard University Press, 1983.

Vahed, K. "The Function of Nuptial Feeding in Insects: A Review of Empirical Studies." *Biological Reviews* 73 (1998): 43–78.

Any Partner Will Do

Brusca, R. C., and G. J. Brusca. *Invertebrates.* Sunderland, Australia: Sinauer Associates, 1990.

Burns, J. M. *BioGraffiti: A Natural Selection.* New York: W. W. Norton, 1975.

Fautin, D. G. "Sexual Stunts of Clownfish." *Natural History* 98 (September 1989): 42–46.

Fischer, E. A. "The Relationship between Mating System and Simultaneous Hermaphro-ditism in the Coral Reef Fish, *Hypoplectrus nigricans* (Serranidae)." *Animal Behaviour* 28 (1980): 620–33.

Fricke, H., and S. Fricke. "Monogamy and Sex Change by Aggressive Dominance in Coral Reef Fish." *Nature* 266 (1977): 830–32.

Hadfield, M. G., and M. Switzer-Dunlap. "Opisthobranchs." In *The Mollusca,* Vol. 7: *Re-production,* ed. A. S. Tompa, N. H. Verdonk, and J. A. M. Van den Biggelaar, pp. 209–350. Orlando, FL: Academic Press, 1984.

The Metamorphoses of Ovid, trans. Mary M. Innes. London: Penguin, 1955.

Michiels, N. K., and B. Bakovski. "Sperm Trading in a Hermaphroditic Flatworm: Reluc-tant Fathers and Sexy Mothers." *Animal Behaviour* 59 (2000): 319–25.

Rollo, C. D., and W. G. Wellington. "Why Slugs Squabble." *Natural History* 86 (September 1977): 46–51.

Sakai, Y., M. Kohda, and T. Kuwamura. "Effect of Changing Harem on Timing of Sex Change in Female Cleaner Fish *Labroides dimidiatus.*" *Animal Behaviour* 62 (2001): 251–57.

Tompa, A. S. "Land Snails (Stylommatophora)." In *The Mollusca*, Vol. 7: *Reproduction*, ed. A. S. Tompa, N. H. Verdonk, and J. A. M. Van den Biggelaar, pp. 47–140. Orlando, FL: Academic Press, 1984.

Warner, R. R. "Mating Behavior and Hermaphroditism in Coral Reef Fishes." *American Scientist* 72 (1984): 128–36.

Survival of the Pampered

Bombeck, E. *if life is a bowl of cherries—what am i doing in the pits?* New York: McGraw-Hill, 1978.

Crump, M. L. "Parental Care." In *Amphibian Biology*. Vol. 2, *Social Behaviour*, ed. H. Heatwole and B. K. Sullivan, pp. 518–67. Chipping Norton, New South Wales: Surrey Beatty & Sons, 1995.

Diesel, R. "Managing the Offspring Environment: Brood Care in the Bromeliad Crab, *Metopaulias depressus*." *Behavioral Ecology and Sociobiology* 30 (1992): 125–34.

Diesel, R., and M. Schuh. "Maternal Care in the Bromeliad Crab *Metopaulias depressus* (Decapoda): Maintaining Oxygen, pH and Calcium Levels Optimal for the Larvae." *Behavioral Ecology and Sociobiology* 32 (1993): 11–15.

Keenleyside, M. H. A. *Diversity and Adaptation in Fish Behaviour*. New York: Springer-Verlag, 1979.

Weygoldt, P. "Complex Brood Care and Reproductive Behavior in Captive Poison-Arrow Frogs, *Dendrobates pumilio* O. Schmidt." *Behavioral Ecology and Sociobiology* 7 (1980): 329–32.

Nests Aren't Just for the Birds

Brockmann, H. J. "Father of the Brood." *Natural History* 97 (July 1988): 33–36.

Coelho, J. "Spurred on to Greater Depths." *Natural History* 111 (June 2002): 20–22.

Fink, L. S. "Costs and Benefits of Maternal Behaviour in the Green Lynx Spider (Oxyopidae, *Peucetia viridans*)." *Animal Behaviour* 34 (1986): 1051–60.

Greene, H. W. *Snakes: The Evolution of Mystery in Nature*. Berkeley: University of California Press, 1997.

Hansell, M. H. *Animal Architecture and Building Behaviour*. London: Longman, 1984.

Lear, E. *A Book of Nonsense*. London: Thomas McLean, 1846.

Mora, G. "Paternal Care in a Neotropical Harvestman, *Zygopachylus albomarginis* (Arachnida, Opiliones: Gonyleptidae)." *Animal Behaviour* 39 (1990): 582–93.

Rosenblatt, J. S., and C. T. Snowdon, eds. *Parental Care: Evolution, Mechanisms, and Adaptive Significance*. Advances in the Study of Behavior. Vol. 25. San Diego, CA: Academic Press, 1996.

Scott, M. P., and J. F. A. Traniello. "Guardians of the Underworld." *Natural History* 98 (June 1989): 32–37.

Van Mierop, L. H. S., and S. M. Barnard. "Further Observations on Thermoregulation in the Brooding Female *Python molurus bivittatus* (Serpentes: Boidae)." *Copeia* (1978): 615–21.

Babies on Board

Crump, M. L. "The Many Ways to Beget a Frog." *Natural History* 86 (January 1977): 38–45.

———. "Parental Care among the Amphibia." In *Parental Care: Evolution, Mechanisms, and Adaptive Significance,* ed. J. S. Rosenblatt and C. T. Snowdon. Advances in the Study of Behavior. Vol. 25, pp. 109–44. San Diego, CA: Academic Press, 1996.

Keegan, H. L. *Scorpions of Medical Importance.* Jackson: University Press of Mississippi, 1980.

Masson, J. M. *The Emperor's Embrace: Reflections on Animal Families and Fatherhood.* New York: Pocket Books, 1999.

Preston-Mafham, K., and R. Preston-Mafham. *The Natural History of Spiders.* Wiltshire, UK: Crowood Press, 1996.

Smith, R. L. "Daddy Water Bugs." *Natural History* 89 (February 1980): 56–63.

A Pouch Full of Miracles

Chia, F.-S. "Brooding Behavior of a Six-Rayed Starfish, *Leptasterias hexactis.*" *Biological Bulletin* 130 (1966): 304–15.

Corben, C. J., G. J. Ingram, and M. J. Tyler. "Gastric Brooding: Unique Form of Parental Care in an Australian Frog." *Science* 186 (1974): 946–47.

Crump, M. *In Search of the Golden Frog.* Chicago: University of Chicago Press, 2000.

Crump, M. L. "Natural History of Darwin's Frog, *Rhinoderma darwinii.*" *Herpetological Natural History* 9 (2002): 21–30.

———. "Parental Care among the Amphibia." In *Parental Care: Evolution, Mechanisms, and Adaptive Significance,* ed. J. S. Rosenblatt and C. T. Snowdon. Advances in the Study of Behavior. Vol. 25, pp. 109–44. San Diego, CA: Academic Press, 1996.

Dawson, T. J. *Kangaroos: Biology of the Largest Marsupials.* Ithaca, NY: Comstock, 1995.

Duellman, W. E., and S. J. Maness. "The Reproductive Behavior of Some Hylid Marsupial Frogs." *Journal of Herpetology* 14 (1980): 213–22.

Masson, J. M. *The Emperor's Embrace: Reflections of Animal Families and Fatherhood.* New York: Pocket Books, 1999.

Menge, B. A. "Brood or Broadcast? The Adaptive Significance of Different Reproductive Strategies in the Two Intertidal Sea Stars *Leptasterias hexactis* and *Pisaster ochraceus.*" *Marine Biology* 31 (1975): 87–100.

Milne, A. A. *The World of Pooh.* New York: E. P. Dutton, 1957.

Moyle, P. B., and J. J. Cech Jr. *Fishes: An Introduction to Ichthyology.* 3rd ed. Upper Saddle River, NJ: Prentice Hall, 1996.

Shuster, S. M. "Changes in Female Anatomy Associated with the Reproductive Moult in *Paracerceis sculpta,* a Semelparous Isopod Crustacean." *Journal of Zoology* (London) 225 (1991): 365–79.

Tyler, M. J., et al. "Inhibition of Gastric Acid Secretion in the Gastric Brooding Frog, *Rheobatrachus silus.*" *Science* 220 (1983): 609–10.

Vincent, A. "A Seahorse Father Makes a Good Mother." *Natural History* 99 (December 1990): 34–42.

Left Holding the Egg

Allport, S. *A Natural History of Parenting.* New York: Harmony Books, 1997.

Clutton-Brock, T. H. *The Evolution of Parental Care.* Princeton, NJ: Princeton University Press, 1991.

Dorst, J. *The Life of Birds.* Vol. 2. London: Weidenfeld and Nicolson, 1971.

Masson, J. M. *The Emperor's Embrace: Reflections of Animal Families and Fatherhood.* New York: Pocket Books, 1999.

Tallamy, D. W. "Insect Parental Care." *BioScience* 34 (1984): 20–24.

Time-Life, eds. *Insects & Spiders.* Alexandria, VA: Time-Life Books, 1992.

Trumbo, S. T. "Parental Care in Invertebrates." In *Parental Care: Evolution, Mechanisms, and Adaptive Significance,* ed. J. S. Rosenblatt and C. T. Snowdon. Advances in the Study of Behavior. Vol. 25, pp. 3–51. San Diego, CA: Academic Press, 1996.

Ustinov, P. *Dear Me.* London: William Heinemann, 1977.

Wilson, E. O. *The Insect Societies.* Cambridge, MA: Belknap Press of Harvard University Press, 1971.

Woolley, T. A. *Acarology: Mites and Human Welfare.* New York: John Wiley & Sons, 1988.

The Life of Birds. BBC Video. David Attenborough. 1998.

Blood Meals: From Bat to Tick Attacks

Askew, R. R. *Parasitic Insects.* New York: American Elsevier, 1971.

Greenhall, A. M., and U. Schmidt, eds. *Natural History of Vampire Bats.* Boca Raton, FL: CRC Press, 1988.

Slansky, F., Jr., and J. G. Rodriguez, eds. *Nutritional Ecology of Insects, Mites, Spiders, and Related Invertebrates.* New York: John Wiley & Sons, 1987.

Stoker, B. *Dracula.* 1897. New York: Bantam Books, 1981.

Earth's Sanitizers

Emlen, D. "Dig It, and They Will Come." *Natural History* 109 (April 2000): 64–69.

Evans, A. V., and C. L. Bellamy. *An Inordinate Fondness for Beetles.* Ontario, Canada: Fitzhenry & Whiteside, 1996.

Evans, G. *The Life of Beetles.* New York: Hafner Press, 1975.

Troyer, K. "Transfer of Fermentative Microbes between Generations in a Herbivorous Lizard." *Science* 216 (1982): 540–42.

Waage, J. K., and G. G. Montgomery. "*Cryptoses choloepi:* A Coprophagous Moth that Lives on a Sloth." *Science* 193 (1976): 157–58.

One Person's Pleasure Is Another Person's Poison

Brown, D. *The Cooking of Scandinavia.* Alexandria, VA: Time-Life Books, 1977.

Coe, S. D. *America's First Cuisines.* Austin: University of Texas Press, 1994.

Hahn, E. *The Cooking of China.* New York: Time-Life Books, 1974.

Livingston, A. D., and H. Livingston. *Edible Plants and Animals: Unusual Foods from Aardvark to Zamia.* New York: Facts on File, 1993.

Livo, L. J., G. McGlathery, and N. J. Livo. *Of Bugs and Beasts: Fact, Folklore, and Activities.* Englewood, CO: Teacher Ideas Press, 1995.

Madsen, D. B. "A Grasshopper in Every Pot." *Natural History* 98 (July 1989): 22–25.

Menzel, P., and F. D'Aluisio. *Man Eating Bugs: The Art and Science of Eating Insects.* Berkeley, CA: Ten Speed Press, 1998.

Post, L. van der. *African Cooking.* New York: Time-Life Books, 1974.

Rombauer, I. S., and M. R. Becker. *Joy of Cooking.* Indianapolis: Bobbs-Merrill, 1972.

Steinberg, R. *Pacific and Southeast Asian Cooking.* Alexandria, VA: Time-Life Books, 1979.

Umstot, M. E. "Eating Snakes in Hong Kong." *Reptile and Amphibian Magazine* 43 (1996): 28–34.

The Incredible Edible Egg

Burton, R. *Bird Behavior.* London: Grenada, 1985.

Crump, M. L., and R. H. Kaplan. "Clutch Energy Partitioning of Tropical Tree Frogs (Hylidae)." *Copeia* (1979): 626–35.

Greene, H. W. *Snakes: The Evolution of Mystery in Nature.* Berkeley: University of California Press, 1997.

Jungfer, K.-H., and P. Weygoldt. "Biparental Care in the Tadpole-Feeding Amazonian Treefrog *Osteocephalus oophagus.*" *Amphibia-Reptilia* 20 (1999): 235–49.

Rombauer, I. S., and M. R. Becker. *Joy of Cooking.* Indianapolis: Bobbs-Merrill, 1972.

Stewart, M. M., and L. L. Woolbright. "Amphibians." In *The Food Web of a Tropical Rain Forest,* ed. by D. P. Reagan and R. B. Waide, pp. 273–320. Chicago: University of Chicago Press, 1996.

Townsend, D. S., M. M. Stewart, and F. H. Pough. "Male Parental Care and Its Adaptive Significance in a Neotropical Frog." *Animal Behaviour* 32 (1984): 421–31.

Life in an Organic Soup

Burns, J. M. *BioGraffiti. A Natural Selection.* New York: W. W. Norton, 1975.

Noble, E. R., and G. A. Noble. *Parasitology: The Biology of Animal Parasites.* Philadelphia: Lea & Febiger, 1964.

Paracer, S., and V. Ahmadjian. *Symbiosis. An Introduction to Biological Associations.* Oxford: Oxford University Press, 2000.

Angling Lures: Masters of Deception

Bond, C. E. *Biology of Fishes.* New York: Saunders College Publishing, 1996.

Greene, H. W., and J. A. Campbell. "Notes on the Use of Caudal Lures by Arboreal Green Pit Vipers." *Herpetologica* 28 (1972): 32–34.

Heatwole, H., and E. Davison. "A Review of Caudal Luring in Snakes with Notes on Its Occurrence in the Saharan Sand Viper, *Cerastes vipera.*" *Herpetologica* 32 (1976): 332–36.

Herald, E. S. *Living Fishes of the World.* New York: Doubleday, 1975.

Maclean, N. *A River Runs through It.* Chicago: University of Chicago Press, 1976.

Murphy, J. B. "Pedal Luring in the Leptodactylid Frog, *Ceratophrys calcarata* Boulenger." *Herpetologica* 32 (1976): 339–41.

Pietsch, T. W. "Dimorphism, Parasitism and Sex: Reproductive Strategies among Deep-sea Ceratoid Anglerfishes." *Copeia* (1976): 781–93.

Radcliffe, C. W., et al. "Observations on Pedal Luring and Pedal Movements in Lepto-dactylid Frogs." *Journal of Herpetology* 20 (1986): 300–306.

Robison, B. H. "Deep-Sea Fishes." *Natural History* 85 (July 1976): 38–45.

Welsh, H. H., Jr., and A. J. Lind. "Evidence of Lingual-Luring by an Aquatic Snake." *Journal of Herpetology* (2000): 67–74.

Stomping for Worms

Conniff, R. *Spineless Wonders: Strange Tales from the Invertebrate World.* New York: Henry Holt, 1996.

Hooker, J. D. "Earthworm Calling." *Wilderness Way Online* 1, no. 3. http://www.wwmag .net/Pages/worm.htm.

Kaufmann, J. H. "Stomping for Earthworms by Wood Turtles, *Clemmys insculpta:* A Newly Discovered Foraging Technique." *Copeia* (1986): 1001–4.

———. "The Wood Turtle Stomp." *Natural History* 98 (August 1989): 8–13.

Kaufmann, J. H., J. H. Harding, and K. N. Brewster. "Worm Stomping by Wood Turtles Revisited." *Bulletin of the Chicago Herpetological Society* 24 (1989): 125–26.

Tinbergen, N. *The Herring Gull's World.* New York: Basic Books, 1960.

Vander Wall, S. B. *Food Hoarding in Animals.* Chicago: University of Chicago Press, 1990.

"World Worm Charming Championships." http://mysite.freeserve.com/wormcharming/.

A Team Effort: How (Some) Ants Get Food

Beebe, W. *Edge of the Jungle.* Garden City, NY: Garden City Publishing, 1926.

Currie, C. R., et al. "Ancient Tripartite Coevolution in the Attine Ant-Microbe Symbio-sis." *Science* 299 (2003): 386–88.

Gotwald, W. H., Jr. *Army Ants: The Biology of Social Predation.* Ithaca, NY: Cornell University Press, 1995.

Hölldobler, B. "The Wonderfully Diverse Ways of the Ant." *National Geographic* (June 1984): 779–813.

Hölldobler, B., and E. O. Wilson. *The Ants.* Cambridge, MA: Belknap Press of Harvard University Press, 1990.

———. *Journey to the Ants: A Story of Scientific Exploration.* Cambridge, MA: Belknap Press of Harvard University Press, 1994.

Hoyt, E. *The Earth Dwellers: Adventures in the Land of Ants.* New York: Simon & Schuster, 1996.

Topoff, H. "Ants on the March." *Natural History* 84 (December 1975): 60–69.

———. "Ant Wars." *Natural History* 96 (January 1987): 62–71.

Chameleons of the Sea

Cousteau, J.-Y., and P. Diolé. *Octopus and Squid: The Soft Intelligence.* Garden City, NY: Doubleday, 1973.

Forsythe, J. W., and R. T. Hanlon. "Foraging and Associated Behavior by *Octopus cyanea* Gray, 1849 on a Coral Atoll, French Polynesia." *Journal of Experimental Marine Biology and Ecology* 209 (1997): 15–31.

Hanlon, R. T., J. W. Forsythe, and D. E. Joneschild. "Crypsis, Conspicuousness, Mimicry and Polyphenism as Antipredator Defences of Foraging Octopuses on Indo-

Pacific Coral Reefs, with a Method of Quantifying Crypsis from Video Tapes." *Biological Journal of the Linnean Society* 66 (1999): 1–22.

Hanlon, R. T., and J. B. Messenger. *Cephalopod Behaviour.* Cambridge: Cambridge University Press, 1996.

Klingel, G. C. *Inagua: An Island Sojourn.* London: Readers Union/Robert Hale, 1944.

Lane, F. W. *Kingdom of the Octopus.* New York: Sheridan House, 1960.

Wells, M. J. *Octopus. Physiology and Behavior of an Advanced Invertebrate.* London: Chapman and Hall, 1978.

Tears of Blood

Manaster, J. *Horned Lizards.* Austin: University of Texas Press, 1997.

Middendorf, G. A., III, and W. C. Sherbrooke. "Canid Elicitation of Blood-Squirting in a Horned Lizard (*Phrynosoma cornutum*)." *Copeia* (1992): 519–27.

Sherbrooke, W. C. *Horned Lizards.* Globe, AZ: Southwest Parks and Monuments Association, 1981.

―――. *Introduction to Horned Lizards of North America.* Berkeley: University of California Press, 2003.

Casting the Insides Outside

Lewis, J. G. E. *The Biology of Centipedes.* Cambridge: Cambridge University Press, 1981.

Mattson, P. *Regeneration.* Indianapolis: Bobbs-Merrill, 1976.

Smith, G. N., Jr. "Regeneration in the Sea Cucumber *Leptosynapta*. I. The Process of Regeneration." In *Regeneration in Lower Vertebrates and Invertebrates.* Vol. 3, pp. 142–52. New York: MSS Information Corp., 1972.

―――. "Regeneration in the Sea Cucumber *Leptosynapta*. II. The Regenerative Capacity." In *Regeneration in Lower Vertebrates and Invertebrates.* Vol. 3, pp. 153–64. New York: MSS Information Corp., 1972.

Smith, L. D. "The Impact of Limb Autotomy on Mate Competition in Blue Crabs *Callinectes sapidus* Rathbun." *Oecologia* 89 (1992): 494–501.

Weimer, B. R. *Nature Smiles in Verse.* Baltimore, MD: Waverly Press, 1940.

Spit 'n' Spray

Blum, M. S. *Chemical Defenses of Arthropods.* New York: Academic Press, 1981.

Brodie, E. D., Jr., and N. J. Smatresk. "The Antipredator Arsenal of Fire Salamanders: Spraying of Secretions from Highly Pressurized Dorsal Skin Glands." *Herpetologica* 46 (1990): 1–7.

Eisner, T. E. "Beetle's Spray Discourages Predators." *Natural History* 75 (1966): 42–47.

Eisner, T. "Chemical Defense against Predation in Arthropods." In *Chemical Ecology,* ed. E. Sondheimer and J. B. Simeone, pp. 157–217. New York: Academic Press, 1970.

―――. "Defensive Spray of a Phasmid Insect." *Science* 148 (1965): 966–68.

―――. *For Love of Insects.* Cambridge, MA: Belknap Press of Harvard University Press, 2003.

Fink, L. S. "Venom Spitting by the Green Lynx Spider, *Peucetia viridans* (Araneae, Oxyopidae)." *Journal of Arachnology* 12 (1984): 372–73.

Greene, H. W. *Snakes: The Evolution of Mystery in Nature.* Berkeley: University of California Press, 1997.

Harmon, D., and A. S. Rubin. *Llamas on the Trail: A Packer's Guide.* Missoula, MT: Mountain Press, 1992.

Livo, L. J., G. McGlathery, and N. J. Livo. *Of Bugs and Beasts: Fact, Folklore, and Activities.* Englewood, CO: Teacher Ideas Press, 1995.

Owen, D. *Camouflage and Mimicry.* Chicago: University of Chicago Press, 1980.

Venom: Serpents' Fangs

Caras, R. A. *Dangerous to Man.* New York: Holt, Rinehart and Winston, 1975.

Greene, H. W. *Snakes: The Evolution of Mystery in Nature.* Berkeley: University of California Press, 1997.

Grenard, S. *Medical Herpetology.* Pottsville, PA: Reptile and Amphibian Magazine, 1994.

Groves, J. D. "Rattlesnake Flags of the American Revolution." *America's First Zoo* 28 (1976): 8.

Minton, S. A., Jr., and M. R. Minton. *Venomous Reptiles.* New York: Charles Scribner's Sons, 1969.

Pough, F. H., et al. *Herpetology.* Upper Saddle River, NJ: Pearson Prentice Hall, 2004.

William Shakespeare: The Complete Works. General editors, S. Wells and G. Taylor. Oxford: Clarendon Press, 1986.

Venom: Centipede and Spider Fangs

Berenbaum, M. R. *Ninety-nine Gnats, Nits, and Nibblers.* Urbana: University of Illinois Press, 1989.

Caras, R. A. *Dangerous to Man.* New York: Holt, Rinehart and Winston, 1975.

Rubio, M. "Is It Time for a Change? An Introduction to Arachnoculture." *Reptile & Amphibian Magazine* 55 (1998): 56–64.

Venom: Spines, Stingers, and Stinging Cells

American Apitherapy Society. www.apitherapy.org.

Caras, R. A. *Dangerous to Man.* New York: Holt, Rinehart and Winston, 1975.

Doyle, A. C. *Sherlock Holmes: The Complete Novels and Stories.* Vol. 2. New York: Bantam Books, 1986.

Hölldobler, B., and E. O. Wilson. *The Ants.* Cambridge, MA: Belknap Press of Harvard University Press, 1990.

Kastner, M., and H. Burroughs. *Alternative Healing.* La Mesa, CA: Halcyon, 1993.

Ohio State University Extension Fact Sheet: Bee and Wasp Stings. 2003. www.ohioline.ag.ohio-state.edu.

Pearson, D. L., and L. Beletsky. *Ecuador and Its Galápagos Islands: The Ecotravellers' Wildlife Guide.* New York: Academic Press, 2000.

Plinius, S. C. (Pliny). *Historia Naturalis. Natural History.* Latin version and English translation by H. Rackham. Vol. 3. Cambridge, MA: Harvard University Press, 1967.

———. *Historia Naturalis. Natural History.* Latin version and English translation by W. H. S. Jones. Vol. 8. Cambridge, MA: Harvard University Press, 1963.

Russell, F. E. *Marine Toxins and Venomous and Poisonous Marine Animals.* Neptune City, NJ: T. F. H. Publications, 1971.

Borrowed and Stolen Defenses

Brodie, E. D., Jr. "Hedgehogs Use Toad Venom in Their Own Defence." *Nature* 268 (1977): 627–28.

International Wildlife Encyclopedia. Vols. 10, 19. New York: Marshall Cavendish Corp., 2002.

Potter, Beatrix. *The Tale of Mrs. Tiggy-Winkle.* New York: Frederick Warne & Co., 1905.

Ross, D. M. "Protection of Hermit Crabs (*Dardanus* spp.) from Octopus by Commensal Sea Anemones (*Calliactis* spp)." *Nature* 230 (1971): 401–2.

Masquerading as Debris

Brandt, M., and D. Mahsberg. "Bugs with a Backpack: The Function of Nymphal Camouflage in the West African Assassin Bugs *Paredocla* and *Acanthaspis* spp." *Animal Behaviour* 63 (2002): 277–84.

Eisner, T., K. Hicks, M. Eisner, and D. S. Robson. "'Wolf-in-Sheep's-Clothing' Strategy of a Predaceous Insect Larva." *Science* 199 (1978): 790–94.

The Fables of Aesop. Retold by F. Barnes-Murphy. Collected and illustrated by R. Barnes-Murphy. New York: Lothrop, Lee & Shepard Books, 1994.

McMahan, E. A. "Adaptations, Feeding Preferences, and Biometrics of a Termite-Baiting Assassin Bug (Hemiptera: Reduviidae)." *Annals of the Entomological Society of America* 76 (1983): 483–86.

———. "Bait-and-Capture Strategy of a Termite-Eating Assassin Bug." *Insectes Sociaux* 29 (1982): 346–51.

———. "Bugs Angle for Termites." *Natural History* 92 (May 1983): 40–47.

Owen, D. *Camouflage and Mimicry.* Chicago: University of Chicago Press, 1980.

Wicksten, M. K. "A Review and a Model of Decorating Behavior in Spider Crabs (Decapoda, Brachyura, Majidae)." *Crustaceana* 64 (1993): 314–25.

Take Cover!

Hoogland, J. L. *The Black-Tailed Prairie Dog: Social Life of a Burrowing Mammal.* Chicago: University of Chicago Press, 1995.

———. "Why Do Gunnison's Prairie Dogs Give Anti-Predator Calls?" *Animal Behaviour* 51 (1996): 871–80.

Kiriazis, J. "Communication and Sociality in Gunnison's Prairie Dogs." PhD diss., Northern Arizona University, 1991.

Macedonia, J. M., and C. S. Evans. "Variation among Mammalian Alarm Call Systems and the Problem of Meaning in Animal Signals." *Ethology* 93 (1993): 177–97.

Seyfarth, R. M., D. L. Cheney, and P. Marler. "Vervet Monkey Alarm Calls: Semantic Communication in a Free-Ranging Primate." *Animal Behaviour* 28 (1980): 1070–94.

Slobodchikoff, C. "The Language of Prairie Dogs." *Plateau Journal* (Museum of Northern Arizona) 6 (2002): 30–38.

Slobodchikoff, C. N., J. Kiriazis, C. Fischer, and E. Creef. "Semantic Information Distinguishing Individual Predators in the Alarm Calls of Gunnison's Prairie Dogs." *Animal Behaviour* 42 (1991): 713–19.

Wilson, E. O. *Sociobiology: The New Synthesis.* Cambridge, MA: Belknap Press of Harvard University Press, 1975.

Living Flashlights

Brusca, R. C., and G. J. Brusca. *Invertebrates.* Sunderland, MA: Sinauer Associates, 1990.

Burgess, G. "Lightning Bug." In *Nature Smiles in Verse,* compiled by B. R. Weimer. Baltimore: Waverly Press, 1940.

Evans, A. V., and C. L. Bellamy. *An Inordinate Fondness for Beetles.* New York: Henry Holt, 1996.

Hanlon, R. T., and J. B. Messenger. *Cephalopod Behaviour.* Cambridge: Cambridge University Press, 1996.

Lloyd, J. E. "Aggressive Mimicry in *Photuris* Fireflies: Signal Repertoires by Femmes Fatales." *Science* 187 (1975): 452–53.

Mathews, R. W., and J. R. Mathews. *Insect Behavior.* New York: John Wiley & Sons, 1978.

McCosker, J. E. "Flashlight Fishes." *Scientific American* 236 (1977): 106–15.

Morin, J. G., et al. "Light for All Reasons: Versatility in the Behavioral Repertoire of the Flashlight Fish." *Science* 190 (1975): 74–76.

Simon, H. *Living Lanterns. Luminescence in Animals.* New York: Viking Press, 1971.

Stolz, U., et al. "Darwinian Natural Selection for Orange Bioluminescent Color in a Jamaican Click Beetle." *Proceedings of the National Academy of Sciences* 100 (2003): 14955–59.

Rapturous and Rapacious Reptiles

Carpenter, C. C. "Communication and Displays of Snakes." *American Zoologist* 17 (1977): 217–23.

Chiszar, D., K. Scudder, H. M. Smith, and C. W. Radcliffe. "Observation of Courtship Behavior in the Western Massasauga (*Sistrurus catenatus tergeminus*)." *Herpetologica* 32 (1976): 337–38.

Gillingham, J. C. "Reproductive Behavior of the Western Fox Snake, *Elaphe v. vulpina* (Baird and Girand)." *Herpetologica* 30 (1974): 309–13.

Jackson, C. G., Jr., and J. D. Davis. "A Quantitative Study of the Courtship Display of the Red-Eared Turtle, *Chrysemys scripta elegans* (Wied)." *Herpetologica* 28 (1972): 58–64.

Kaufmann, J. H. "The Social Behavior of Wood Turtles, *Clemmys insculpta,* in Central Pennsylvania." *Herpetological Monographs* 6 (1992): 1–25.

Lewke, R. E. "Neck-Biting and Other Aspects of Reproductive Biology of the Yuma Kingsnake (*Lampropeltis getulus*)." *Herpetologica* 35 (1979): 154–57.

Nash, Ogden. *I Wouldn't Have Missed It: Selected Poems of Ogden Nash,* ed. L. Smith and I. Eberstadt. Boston: Little, Brown, 1975.

Pough, F. H., et al. *Herpetology.* 3rd ed. Upper Saddle River, NJ: Pearson Prentice Hall, 2004.

Smelling Is Understanding

Ali, M. F., and E. D. Morgan. "Chemical Communication in Insect Communities: A Guide to Insect Pheromones with Special Emphasis on Social Insects." *Biological Reviews* 65 (1990): 227–47.

Fitzgerald, T. D. "Caterpillar on a Silken Thread." *Natural History* 92 (February 1983): 56–63.

Fitzgerald, T. D., and S. C. Peterson. "Elective Recruitment Communication by the Eastern Tent Caterpillar (*Malacosoma americanum*)." *Animal Behaviour* 31 (1983): 417–23.

Free, J. B. *Pheromones of Social Bees.* Ithaca, NY: Cornell University Press, 1987.

Hölldobler, B., and E. O. Wilson. *The Ants.* Cambridge, MA: Belknap Press of Harvard University Press, 1990.

Stoddart, D. M. *Mammalian Odours and Pheromones.* The Institute of Biology's Studies in Biology no. 73. London: Edward Arnold Ltd., 1976.

White, A. M., R. R. Swaisgood, and H. Zhang. "The Highs and Lows of Chemical Communication in Giant Pandas (*Ailuropoda melanoleuca*): Effect of Scent Deposition Height on Signal Discrimination." *Behavioral Ecology and Sociobiology* 51 (2002): 519–29.

Wiley, F. A., ed. *Ernest Thompson Seton's America.* Garden City, NY: Doubleday, 1963.

Love Potion Number Nine

Garstka, W. R., B. Camazine, and D. Crews. "Interactions of Behavior and Physiology during the Annual Reproductive Cycle of the Red-Sided Garter Snake (*Thamnophis sirtalis parietalis*)." *Herpetologica* 38 (1982): 104–23.

Gerardi, M. H., and J. K. Grimm. *The History, Biology, Damage, and Control of the Gypsy Moth* Porthetria Dispar (L.). Cranbury, NJ: Associated University Presses, 1979.

Hays, W. S. T. "Human Pheromones: Have They Been Demonstrated?" *Behavioral Ecology and Sociobiology* 54 (2003): 89–97.

Lemaster, M. P., and R. T. Mason. "Variation in a Female Sexual Attractiveness Pheromone Controls Male Mate Choice in Garter Snakes." *Journal of Chemical Ecology* 28 (2002): 1269–85.

Mendonca, M. T., and D. Crews. "Control of Attractivity and Receptivity in Female Red-Sided Garter Snakes." *Hormones and Behavior* 40 (2001): 43–50.

Pough, F. H., et al. *Herpetology.* 3rd ed. Upper Saddle River, NJ: Prentice Hall, 2004.

Shoumatoff, A. "The Greatest Show on Earth." *Audubon* 106 (2004): 32–37.

Voigt, C. C. "Individual Variation in Perfume Blending in Male Greater Sac-Winged Bats." *Animal Behaviour* 63 (2002): 907–13.

Voigt, C. C., and O. von Helversen. "Storage and Display of Odour by Male *Saccopteryx bilineata* (Chiroptera, Emballonuridae)." *Behavioral Ecology and Sociobiology* 47 (1999): 29–40.

It Pays to Be Neighborly

Bradbury, J. W., and S. L. Vehrencamp. *Principles of Animal Communication.* Sunderland, MA: Sinauer Associates, 1998.

Frost, Robert. *The Poetry of Robert Frost,* ed. E. C. Lathem. New York: Henry Holt, 1979.

Guffey, C., J. G. MaKinster, and R. G. Jaeger. "Familiarity Affects Interactions between Potentially Courting Territorial Salamanders." *Copeia* (1998): 205–8.

Jaeger, R. G. "Dear Enemy Recognition and the Costs of Aggression between Salamanders." *American Naturalist* 117 (1981): 962–74.

Rosell, F., and T. Bjorkoyli. "A Test of the Dear Enemy Phenomenon in the Eurasian Beaver." *Animal Behaviour* 63 (2002): 1073–78.

Temeles, E. J. "The Role of Neighbours in Territorial Systems: When Are They 'Dear Enemies'?" *Animal Behaviour* 47 (1994): 339–50.

Weary, D. M., R. E. Lemon, and E. M. Date. "Neighbour-Stranger Discrimination by Song in the Veery, a Species with Song Repertoires." *Canadian Journal of Zoology* 65 (1987): 1206–9.

Whiting, M. J. "When to Be Neighbourly: Differential Agonistic Responses in the Lizard *Platysaurus broadleyi.*" *Behavioral Ecology and Sociobiology* 46 (1999): 210–14.

Index